Exploring Atmospheres Ethnographically

The notion of atmosphere has always been part of academic discourse, but often refers to something vague and diffuse – a phenomenon connected with our affective engagement with the world that is difficult to grasp. This volume develops and refines the concept of atmosphere, seeking to render it productive for anthropological and social scientific research by bringing together a range of original ethnographic studies in combination with thorough investigation of the use of the term in language.

With chapters that examine dimensions of atmosphere through topics of interdisciplinary concern, such as learning and the acquisition of skills, the experience of place, affect and mood, and the perception of weather and environment – whether in natural landscapes, medical and educational settings, homes or creative contexts – *Exploring Atmospheres Ethnographically* analyses the relational and transformational processes through which people perceive, experience and live in a moving atmospheric world. As such, it will appeal to scholars of anthropology, sociology and cultural studies with interests in space and place, sensory ethnography and affect.

Sara Asu Schroer is a Research Fellow in the department of anthropology at the University of Aberdeen, UK.

Susanne B. Schmitt is a transdisciplinary maker, anthropologist and creative director of "How to Not be a Stuffed Animal".

Anthropological Studies of Creativity and Perception
Series Editor: Tim Ingold

The books in this series explore the relations, in human social and cultural life, between perception, creativity and skill. Their common aim is to move beyond established approaches in anthropology and material culture studies that treat the inhabited world as a repository of complete objects, already present and available for analysis. Instead these works focus on the creative processes that continually bring these objects into being, along with the persons in whose lives they are entangled. All creative activities entail movement or gesture, and the books in this series are particularly concerned to understand the relations between these creative movements and the inscriptions they yield. Likewise, in considering the histories of artefacts, these studies foreground the skills of their makers-cum-users, and the transformations that ensue, rather than tracking their incorporation as finished objects within networks of interpersonal relations. This series is interdisciplinary in orientation, with the concern of the titles always being with the practice of interdisciplinarity: on ways of doing anthropology with other disciplines, rather than doing an anthropology of these subjects. Through this anthropology with focus, they aim to achieve an understanding that is at once holistic and processual, dedicated not so much to the achievement of a final synthesis as to opening up lines of inquiry.

Titles in series

Reflections on Imagination
Human Capacity and Ethnographic Method
Edited by Mark Harris and Nigel Rapport

Craftwork as Problem Solving
Ethnographic Studies of Design and Making
Edited by Trevor H. J. Marchand

Exploring Atmospheres Ethnographically
Edited by Sara Asu Schroer and Susanne B. Schmitt

For a full list of titles please visit:

www.routledge.com/Anthropological-Studies-of-Creativity-and-Perception/book-series/ASHSER1315

Exploring Atmospheres Ethnographically

Edited by Sara Asu Schroer and Susanne B. Schmitt

Routledge
Taylor & Francis Group
LONDON AND NEW YORK

First published 2018 by Routledge

2 Park Square, Milton Park, Abingdon, Oxfordshire OX14 4RN
52 Vanderbilt Avenue, New York, NY 10017

Routledge is an imprint of the Taylor & Francis Group, an informa business

First issued in paperback 2020

British Library Cataloguing-in-Publication Data
A catalogue record for this book is available from the British Library

Library of Congress Cataloging-in-Publication Data
A catalog record for this book has been requested

ISBN: 978-1-4724-6833-8 (hbk)
ISBN: 978-0-367-59366-7 (pbk)

Typeset in Times New Roman
by Apex CoVantage, LLC

Contents

Figures

Contributors

Julia Bee is assistant professor for image theory at Bauhaus Universität Weimar. She works on forms of visual knowledge and ethnographic film. Articles include: "Dramatization of Beginnings. The Intros of Homeland, True Blood and True Detective" (2016), "Violence, Power and Desire. Toward a Politics of Perception" (2015) and "Experience-Images and Fabulations: The Archive of Visual Anthropology" (forthcoming).

Gerko Egert is a postdoctoral researcher at the Institute for Applied Theater Studies, Justus-Liebig-Universität, Gießen, where he works on a project entitled "Choreopower. Choreography and the Politics of Movement". His research deals with contemporary dance, human and nonhuman choreographies, politics of movement, weather and touch. Publications include "Choreographing the Weather – Weathering Choreography" (TDR 2016).

Caroline Gatt is a research fellow (Knowing from the Inside) at the University of Aberdeen. Her recently published book is entitled *An Ethnography of Global Environmentalism: Becoming Friends of the Earth* (2017, Routledge). From 2001 to date she also carried out training and research in laboratory theatre, with groups in Malta, Italy and the UK.

Paolo Gruppuso is Honorary Research Fellow at the Department of Anthropology of the University of Aberdeen. He is interested in environmental conservation, landscape, agriculture, sustainability and political ecology. He has conducted research on topics including environmental conflicts, water and wetlands management, environmental history and environmental education.

Todd E. Nicewonger is Project Director for Destination Areas at Virginia Tech. He received his Ph.D. in Applied Anthropology from Columbia University, Teachers College and is currently conducting research on trans-disciplinary research and educational practices.

Friedlind Riedel is a doctoral fellow at the Competence Centre for Media-Anthropology at Bauhaus University Weimar. As a musicologist and anthropologist, her interests lie with cultural histories of listening in Southeast Asia and with the peculiar relationship between music and mood. She has conducted extensive research on rituals in Myanmar and is currently working on a book that enquires into music and listening at the threshold of the human.

Susanne B. Schmitt is an anthropologist and interdisciplinary artist. Her work focuses on creative collaboration within and beyond the label of "art meets science", multispecies worlds, and the aesthetic dimensions of the workplace. She holds a Ph.D. from the University of Munich based on an ethnography of a German Medical Museum as a viscerally experienced workplace. She is creative director of "How to Not be a Stuffed Animal" (together with Laurie Young, funded by VW Foundation). Affiliations include the Sense Lab in Montreal and the Rachel Carson Center for Environmental Humanities in Munich.

Sara Asu Schroer holds a research fellowship at the interdisciplinary ERC project *Arctic Domus* at the department of anthropology, University of Aberdeen. In her current research she is combining approaches from environmental anthropology, multi-species ethnography and Science and Technology Studies to explore the relationships between humans and birds of prey in falconry, captive breeding and conservation. She is convenor of the EASA network *Humans and Other Living Beings*.

Robert Shaw is a lecturer in geography at Newcastle University. His research explores the social and geographical dimensions of night. He is published in various social science journals and is currently writing a book titled *The Nocturnal City* due in 2018.

Anette Stenslund is a sociologist currently lecturing in cultural sociology at the Department of Sociology, University of Copenhagen. She holds a Ph.D. in medicine, culture and society and her research interests include atmosphere, affect, aesthetics of engagement and sensory studies.

Stefan Wellgraf studied social and cultural studies in Berlin, Frankfurt/Oder, Paris and New York. His dissertation research, "Hauptschüler. Zur gesellschaftlichen Produktion von Verachtung", was published in 2012. He currently works at the European University Viadrina in Frankfurt/Oder.

Andrew Whitehouse is an environmental anthropologist and birder working at the University of Aberdeen. He has conducted research into conservation issues in Scotland and more recently has investigated people's relations with birds through sound as part of the *Listening to Birds Project*. As well as considering the role of bird sounds in people's sense of place, time and season, he examined human interactions with birds through sound and practices of skilled listening in sound identification. He is also co-editor of the volume *Landscapes Beyond Land: Routes, Aesthetics, Narratives*.

Acknowledgements

This volume began at a panel entitled "Exploring atmospheres: an anthropological approach?", which took place at the Association of Social Anthropologists (ASA) conference held in Edinburgh, UK in 2014. We are grateful to the organisers of this conference for providing the opportunity to hold the panel, as well as to the many people, in addition to the presenters themselves, who attended and contributed to what turned out to be a lively and stimulating panel. We also thank Tim Ingold for his role as discussant and for later on supporting us in producing this edited volume. We further would like to acknowledge the constructive criticism made by an anonymous reviewer appointed by the publisher. We especially want to thank the many fieldwork interlocutors whose cooperation in the various research projects on which the presented ethnographic material is based on made this volume possible. Finally, we thank the contributors to this volume for their enthusiasm and patience through out the production of the volume. Sara Asu Schroer also wants to acknowledge the funding of the ERC that is currently funding her research fellowship as part of the *Arctic Domus* project at the University of Aberdeen. Susanne B. Schmitt is grateful for the Volkswagen Foundation's Art and Science in Motion grant.

1 Introduction

Thinking through atmospheres

Sara Asu Schroer and Susanne B. Schmitt

Introduction

Atmospheres. Like clouds in the sky they are ever forming and reforming, appearing and disappearing, never finished or at rest. Atmospheres can be sensed by a singular subject yet have collective affective qualities that evade the singular; they can be created but are also co-creating the ways through which we sense and perceive in the world. The notion of atmosphere has always been present as an undercurrent in social anthropology. It often occurs in texts and conversations, as part of ethnographic descriptions and personal field notes. In these instances, atmospheres seem to refer to a phenomenon that stems from our affective engagement with the world – evocative and difficult to grasp in terms of rational explanation. In this volume we are interested in questions of how atmospheres might be addressed ethnographically, leading us to further rethink the boundaries between the material and immaterial, presence and absence, individual and collective as well as body and place.

Academic interest in understanding atmospheres has gained momentum in recent years, particularly in philosophy (Böhme 1995; Schmitz et al. 2011; Griffero 2014; Schmitz 2014), affect and non-representational theory (Anderson 2009; Stewart 2011; Anderson and Ash 2015) and studies of urban and architectural contexts (Zumthor 2006; Böhme 2013; Bille and Sørensen 2016). Here atmospheres are usually understood as a crucial part of human life – both individual and collective – that influences identities, experiences and relationships. Most work to date, however, has focussed more on the philosophy of atmosphere as a concept, rather than on ethnographic enquiry into its various manifestations in social life. *Exploring Atmospheres Ethnographically* seeks to investigate the force and meaning of atmosphere in everyday life. Its contributions analyse the relational and transformational processes through which humans and other living beings perceive, experience and live together in a moving atmospheric world. The chapters, written from an ethnographic and anthropological perspective, examine dimensions of atmosphere through topics of interdisciplinary concern, such as learning and the acquisition of skills, the experience of place, affect and mood, and the perception of weather and environment – be it in landscapes, medical and educational settings, homes or creative contexts.

Most of the chapters were first presented at a panel at the 2014 conference of the Association of Social Anthropologist (ASA) in Edinburgh where we gathered to explore the notion of atmosphere and its potential for anthropological research, a call that was met with much interest and enthusiasm. Other contributions have found their way into the collection at a later stage. When we put together the contributions for the initial panel, it became obvious that although the notion of atmosphere was gaining momentum in anthropology – or maybe we should say it was "in the air" – it was also a field of ideas nourished from diverse sources: German phenomenology, affect studies and non-representational theory, as well as by a renewed interest in rethinking materiality in terms of its emergent and transformative qualities. These influences, however, as they figured in the contributions, converged on similar problems and questions: Are atmospheres media or objects of perception, metaphors or material phenomena? Are they representations, and are they representable? Can they be created, and what do they do? How can atmospheres be affects and effects at the same time? How can we think about atmospheres through air and other elements and substances?

These questions were tied to sites and situations across different ethnographic contexts, interactions and even species. The result is a variety of anthropological case studies that approach the phenomenon of atmospheres from the ground of ethnography. Each contributor shows in pivotal yet not always in consonant ways, how atmospheres matter in the social settings they studied. Retrospectively, we see the contributions gathering along three emerging threads. The first thread centres on an exploration of the atmospheres of institutions and places. Our authors explore the overwhelming feeling of boredom in schools for socio-economically disadvantaged students; look at how night-time economies in British cities use atmospheric strategies to create the ideal pub experience; and unearth the miasmic past and atmospheric present of Italian wetlands. A second thread running through the chapters is concerned with more-than-human atmospheres. Through the ethnographic lens of falconry, aquarium design, and the meaning of birdsong for place making in Britain, Australia and New Zealand, the authors trace how atmospheres are pivotal for conviviality with other living beings, whether they are mollusca or falcons. Following the third thread, chapters on art, music, and creative production discover atmosphere as a pedagogical design tool for fashion design students in Antwerp, explore the role of breath and air in experimental theatre, look at how artistic interventions create a hospital atmosphere through minimal gesture and shift our attention to the process of multisensory, atmospheric events aimed at audiences, whether they are in Gamelan music-making in Southern China or ethnographic filmmaking in the North Atlantic fishery.

Atmospheres and anthropology?

Whilst this volume is novel in explicitly foregrounding atmospheres in ethnographic research, the topic is by no means a novel one for anthropologists to write and think about, as we show below with brief examples from the writings of Emile Durkheim, Marcel Mauss and Clifford Geertz. In fact, phenomena of atmosphere – also addressed using related terms such as ambience, mood, presence, aura or tone – have from early on been crucial if not always central to the anthropological enterprise, as objects of study, as aspects of lived experience during fieldwork, and as creations of ethnographic writing, museum exhibitions and film.

In *The Elementary Forms of the Religious Life* ([1912] 2001), for instance, Durkheim observes the powerful effect of collective feelings in religious life, co-created through the execution of religious rituals and ceremonies. He argues that participants experience the atmosphere of religious events as an "electricity" that induces collective emotional excitement and "delirium". In these moments of "collective effervescence", every participant is supposed to experience an extraordinary and impersonal force or energy that has the power to raise them to an ideal realm (Durkheim 1912: 162, 171). This for him is a central unifying force that enables individuals to identify strongly and feel solidarity with their social group (Morris 2003: 120). Durkheim, then, identified atmosphere as central to the workings of society (Debaene 2014), comprising a domain of affectivity at once created by individuals, whilst yet experienced as lying beyond them, something powerful yet without clear-cut contours. This double-edged nature of the atmosphere as at once interior and exterior to the individual also pervades Durkheim's understanding of what he called social facts, "manners of acting, thinking, and feeling external to the individual, which are invested with a coercive power by virtue of which they exercise control over him" (Durkheim 2001: 21).

Whilst for Durkheim social facts could be studied objectively as objects of scientific research, his student Marcel Mauss, in his essay *The Gift* ([1925] 2002), introduced an important innovation which was to link the study of the total social fact to the lived experience of the ethnographer (Debaene 2014). In so doing he acknowledged ethnographic immersion and personal experience as a central aspect of anthropological knowledge production. Mauss actually characterised the gift – the total social fact par excellence – as an 'atmosphere' permeating social life: "a considerable part of our morality and our lives themselves are still permeated with this same atmosphere of the gift, where obligation and liberty intermingle" (Mauss 2002: 83). An even more explicit use of the term can be found in his *Manual of Ethnography* ([1926] 2007) in which he highlights the centrality of lived experience to anthropological research: "one will be able to define the moral tone of the society under study, making an effort to remain within the ethos [*atmosphère*] of the society: It is good to practice the vendetta, it is good to be able to offer a human head to your fiancée" (Mauss, cited in Debaene 2014: 73).

Atmosphere in Mauss' conception is more than an aspect of aesthetic style that, as ornamentation, could be easily foregone. Rather the ethnographer's grasp of the atmosphere of the people she is working with is evidence of her holistic understanding of their lives. To grasp the atmosphere of a ritual, gathering or economic exchange means achieving a holistic understanding of the phenomenon, rather than robbing it of its liveliness. It is with this in mind that we are alerted to how "Mauss's famous remark 'We touch upon fundamentals' occurs at the moment when a society stops being perceived in abstract terms and appears instead as 'the feeling of men, in their minds and in flesh and blood'". Here atmosphere, as a total social fact, becomes the object of study in itself, which for Mauss can be recognised "in the way it 'permeates' individual behaviours" (Debaene 2014: 73).

The notion of atmospheric permeation is also evident in Anglophone literature, often described with the term 'ethos' first introduced to anthropology by Ruth Benedict (1934), and developed by Gregory Bateson (1936) and Clifford Geertz (1973) (see also Nuckolls 1995). In *The Interpretation of Cultures*, for example, Geertz states: "A people's ethos is the tone, character, and quality of their life, its moral and aesthetic style and mood. It is the underlying attitude towards themselves and their world" (1973: 127). It is constituted as a set of moods and motivations; indeed Geertz describes moods in atmospheric terms: springing "from certain circumstances . . . they are responsive to no ends. Like fogs, they just settle and lift; like scents, suffuse and evaporate" (1973: 97).

Evidently, the phenomena of atmosphere have long been recognised as central to understanding human life and meaning-making. However, atmosphere has not become a focus of anthropological inquiry until now. As Vincent Debaene suggests with regard to Mauss, the reason might be that the term was considered at the time too vague and general to be worth pursuing (Debaene 2014: 75–76). The experience of atmosphere in the field seemed to conflict with the positivist conventions of scientific writing and was difficult to fit into modern categories. Current critiques of earlier conceptions have however led to increased interest in relational and process-oriented approaches, and in many research fields these approaches have become commonplace. In this changed climate a refocusing on atmosphere, understood as lying at the intersection of spatial, sensory, material and affective dimensions of social life, seems pertinent. It requires us to rethink the relationships between subject and object, self and world, perceiver and the perceived.

Current interest in atmospheres

Whilst most of the literature to date has been on the philosophy of atmosphere and aesthetic discourse (Böhme 1993; Griffero 2014), fewer studies look at atmospheres as they are socially and materially produced (Bille 2015; Pink and Leder Mackley 2016, are two exceptions). Bille and his co-authors, for instance, look at how public and private spaces also entail the possibility of staging atmospheres deliberately. Amongst other topics they investigate the co-production of football game atmospheres in stadia by fans and clubs, the sensory and emotional production of urban territories, the making of *hygge* (cosy) atmospheres in Denmark and the possibilities for approaching atmosphere in the remote past. In another collection Edensor and Sumartojo (2015) explore, in a similar vein, the possibilities of designing atmospheres purposefully through the arrangement and orchestration of materials, things, people and places.

Current discourse on atmosphere in the Anglophone literature draws heavily on the little that has been translated of the work of German philosopher Gernot Böhme. For Böhme atmospheres are intermediate phenomena that lie at the intersection of the subjective and the objective:

> Atmospheres are indeterminate above all as regards their ontological status. We are not sure whether we should attribute them to the objects or

> environments from which they proceed or to the subjects who experience them. We are also unsure where they are. They seem to fill the space with a certain tone or feeling like a haze.
>
> (Böhme 1993: 114)

Atmospheres, for Böhme, are neither to be attributed to the experiencing subject alone, nor to be regarded as belonging to the physical environment. He understands them as both experientially and conceptually ambiguous, at once material and immaterial, subjective and objective. Key to Böhme's conception of atmosphere are what he terms the ecstasies of things, that is the ways in which things are sensuously and qualitatively present in particular situations. Atmospheres arise out of the constellation of things and people and are understood as the perceived quality of a certain situation (for a critique of Böhme and an elaboration on the work of phenomenologist Hermann Schmitz, see Riedel 2015 and this volume). Böhme's approach represents the legacy of philosophical thinking on atmosphere, and of related concepts such as *ambience* and *Stimmung*, that has come down to us from Franco-German phenomenology, and that conceives of the atmospheric as both an intermediary of human experience and pre-reflexive (Heidegger [1927] 1996; Merleau-Ponty [1964] 1968; see also Throop 2014).

Aside from phenomenological sources the notion of atmosphere is often linked to the study of affect. This emphasis is particularly evident in human geography and non-representational theory (Anderson and Ash 2015). For Anderson, for instance, atmosphere helps him investigate the differences between affect and emotion (Anderson 2009). One dominant approach to affect in the social sciences has been in the work of philosopher Brian Massumi (2002), who suggests that affect lies beyond individual consciousness and discourse as well as beyond emotional and bodily responses. Following Deleuze, affect is understood in terms of "intensities", understood as a "set of flows moving through the bodies of human and other beings" (Thrift, cited in Bille et al. 2014: 34). As Bille and his colleagues suggest, within these discourses the notion of atmosphere itself is often taken for granted and used synonymously with affect (Bille et al. 2014: 35). This, however, risks silencing several other dimensions of how atmospheres are constituted, the affective dimension being only one of them. This is especially challenging for studies interested in the social and material creation of the atmospheric as it would suggest that atmospheres pre-exist the people who experience them, hence portraying people and other living beings as passively moved rather than as also actively involved in constituting atmospheres (see also Edensor and Sumartojo 2015)

Moving beyond a narrow focus on the affective constitution of the atmospheric, several authors have drawn attention to apparently intangible aspects of life such as lights, colours, odours, temperature, air quality and more generally the weather, in and through which people perceive the world. Thibaud, for instance, emphasises that to understand the social fabric of urban environments it is crucial to focus on elements rather than on objects (Thibaud 2015). This interest is reflected in a variety of publications, across many disciplines, aimed at rethinking materiality and its ontological status; shifting from a dichotomous understanding of matter and life to a more dynamic understanding in which materiality is seen to be in a continuous process of emergence (for an overview see Harvey 2013). Matter, in this understanding, is not contained by boundaries that render it inert but rather surfaces between states of being assumed to be porous, allowing for variability and transformation from states of fixity to suspension (Ingold 2007; Anderson and Wylie 2009). In this material reading, atmosphere becomes more than a metaphor and instead part of the fabric of human and nonhuman ways of living. In anthropology this aspect has recently been approached by Tim Ingold (2012) who is interested in bringing together the aesthetic notion of atmosphere with the atmospheric phenomenon of weather. His main point is to show that atmospheres are not, as often described, vague affective phenomena, but are in fact tied up with meteorological conditions that are also in turn sensual and affective. In contrast to Böhme's insistence of locating atmosphere in the interstitial spaces in-between objects, Ingold highlights its permeating and immersive characteristics. Focusing on the experience of weather, wind and light, he understands atmosphere as a medium, an "all enveloping experience" (Ingold 2011: 134; see also 2015: 73–78) in and through which people perceive.

Exploring atmospheres ethnographically

The chapters in this volume engage with the themes sketched out above, whilst also contributing novel and refreshing perspectives by allowing the notion of atmosphere, situations, stories and histories to unfold ethnographically. Throughout most chapters, the ethnographic texts and interpretations reveal atmosphere to be a more-than-human phenomenon co-created by the moods and practices of living beings, the forces of weather, light, water and air, as well as by landscapes, architectures, artworks, music, sound and colours. With this embrace of the more-than-human and ecological dimensions of atmosphere the contributors expand the current literature and thus connect to a larger anthropological rethinking of sociality and materiality (e.g. Tsing 2012, 2013). Furthermore, whilst finding inspiration in phenomenological approaches, some of the contributions are also critical of their sometimes universalising and ahistorical tendencies, thus highlighting their limits for an ethnographically based understanding of atmosphere that should be sensitive to the partial, situational and historical aspects of human and nonhuman lifeworlds.

Atmospheres, as Marcel Mauss observed, require a holistic understanding based in ethnographic encounters. More than any other concept, atmosphere helps us to grasp the ethnographer's position as both a maker and perceiver, (mis)interpreter and co-creator of atmospheric worlds. For ethnographers, writing itself becomes a pathway into the atmospherics of lively situations and events. It is a way of both making accessible and of attuning oneself to aspects of human and more-than-human existence (Stewart 2011). Doing and writing thick and vivid ethnography, then, involves cultivating sensitivity and attention – to the smell of a corridor, to birdsong, to the decaying structures of a school building or the mood of a falcon as the weather changes. It asks for "attention to the matterings, the complex emergent worlds, happening in everyday life" (Stewart 2011: 452). The contributors to this volume have presented their work through such writing, approaching the topic of atmosphere from the mists of their ethnographic material and through the lens of different perspectives.

In the second chapter Stefan Wellgraf seeks to relate the ruination of infrastructures and the everyday experiences of students at a Berlin Hauptschule, a school attended by the less privileged children of a neighbourhood that finds itself in a downward spiral of socio-economic decay. Drawing on ethnographic fieldwork in the everyday happenings at the school, Wellgraf offers a political reading of the emergence of what he calls "atmospheres of boredom" as they are shaped by historical, material and political processes and practices. Socio-political neglect, material ruination and the stark routines of schooldays contribute to students' alienation through creating an all-encompassing atmosphere tied to a futureless present.

The third chapter, by Robert Shaw, explores the atmosphere of northeast English pubs by drawing on his extended fieldwork in the night-time-economy, and on George Orwell's fictional account of *The Moon under Water*. "If you are asked why you favour a particular public house, it would seem natural to put the beer first, but the thing that most appeals to me about *The Moon Under Water* is what people call its 'atmosphere'" (Orwell 1946). Drawing on non-representational approaches, Shaw shows how ethnography allows us to investigate the multi-layered, sensory, affective and social dimensions of atmospheres. Interested in the production and manipulation of bar spaces, he understands bar space as formed through both material and affective forces that combine to create increasingly consumption-oriented realms of leisure.

Vapours, air and smoke are the elementary forms of atmosphere adrift in chapter four, in which Paolo Gruppuso explores the composition of Italian marshlands, wet landscapes known historically as emanating miasmatic forces of bad air. In the Italian Pontine marshlands, bodies are porous and landscapes are vaporous. Atmosphere here materialises in the disease of malaria, literally 'bad air', and in columns of smoke rising from burn-beating practices in the bog, which connect air and earth. Drawing together historical accounts of the marshes and fieldwork in contemporary Agro Pontino, Gruppuso entwines aesthetic and meteorological notions of atmosphere. He demonstrates vividly the sensory and material presence of atmospheres that permeate and shape the lives of those living and working in the marshes. Atmosphere is here understood not as a medium of perception but rather as a phenomenon to be encountered in the marshes – perceived as terrifying and beautiful at the same time.

Andrew Whitehouse, in Chapter five, moves us away from 'bad air' into air filled by birdsong. How does birdsong create a sense of place and being and how is this related to thinking about atmospheres? This question is explored through narratives of birdsong enthusiasts in Britain, Australia and New Zealand. Whitehouse considers the homely atmospheres created by the songs of familiar species of birds as well as the sensations of being out of place that emerge when memories, sense of place and the atmospheric whole evoked by birdsong clash. Focusing on sound and the act of listening to birds he understands the everyday encounters of humans and birds as embedded in shifting atmospheres, ongoing events, inevitably more-than-human and more-than-subjective.

Birds also take centre stage in the Chapter 6, in which Sara Asu Schroer explores how beings with different perceptual abilities learn to attune to each other and their environments. She explores this issue through detailed ethnographic observations of the taming and training of falconry birds that will eventually lead to a cooperative hunting companionship. The airborne constitution of birds draws our attention to the atmospheric forces of the weather that powerfully influence their lines of flight as much as the movements of the earthbound humans. The human-bird encounter, moreover, requires an attention to the moods of birds and humans that are experienced as spatially tangible as well as reaching beyond the subjective experience of the individual. Atmosphere here appears as dynamic, temporal milieux in and through which humans and birds of prey learn to communicate with each other.

From the currents of the weather we dive into the element of water. In Chapter 7 Susanne B. Schmitt traces the making of aquariums, their visitors and the manifold lifeworlds they contain. Aquariums are early and crucial sites for ecological experimentation and knowledge-making that simultaneously carry a great deal of aesthetic appeal. Schmitt's discussion of aquarium atmospheres engages with the 'how' of the creation of underwater worlds across atmospheric substances like water, air and light. To capture the range of aesthetic and ecological decisions that inform the making of aquariums, she coins the notion of *charismatic ecologies* and demonstrates their more-than-human constitution by drawing on ethnography with an aquarium design company, curators of public aquariums as well as historical records.

Chapter 8, written by Lisa Bee and Gerko Egert, tackles the crucial role of atmosphere in ethnographic filmmaking in an equally aquatic setting. Drawing on William James' notion of experience as chance, flux, stream and becoming, they offer an experiential viewing of the film *Leviathan* as colour-weather-worlding. *Leviathan*, directed by Lucien Castaing-Taylor and Véréna Paravel of the Harvard Sensory Ethnography Lab, follows the work of the North American fishing industry on a fishing trawler. Taking an approach that de-centres the human perspective, the authors argue that atmosphere is an event co-created by the manifold movements and perspectives of human and nonhuman agents. Atmosphere here is not understood as a medium or as an additive, connecting already established subject and object positions. It is rather understood as a process of differential becoming that creates and combines movements as non-linear chains of occasions of experience and movement.

Todd Nicewonger investigates another form of creative making in chapter nine. He uses his ethnography of fashion design pedagogy in Antwerp as a springboard from which to analyse atmosphere as key to understanding design processes of making and imagining. In the design school, atmosphere has emerged as an important pedagogical idiom by which aesthetic innovation is described and socially patterned. Building on ethnographic observation and participation in the everyday engagements of students and tutors, Nicewonger shows how affect is the connective tissue that intimately binds the social learning process of design practice and aesthetic decision-making. In the working styles and pedagogies of fashion designers that tie together social and aesthetic phenomena, he also finds valuable inspiration for ethnography.

Can an actor deliberately generate an atmosphere? Should we consider an atmosphere a medium? How can we know an atmosphere? Caroline Gatt poses these questions in Chapter 10 in relation to the craft of performance through the ethnographic lens of working with experimental theatre makers. She emphasises that shaping an atmosphere is central to the actor's work and as such offers insight into how we know atmospheres. Gatt suggests, based on her ethnographic material on the actor's work on vibrational song and Taijiquan, that attentional practices, the airs, grounds and other constituents of the places we move through and live in are partners in the generation of atmospheres. She argues that it is one's participation in the very creation of those atmospheres that enables knowledge of them.

Why does a scentless installation by the Scandinavian artist duo Elmgreen & Dragset, depicting an institutional corridor, smell like a hospital? In Chapter 11 Anette Stenslund approaches this curious relation, observed in an art gallery in Denmark, through the theoretical figure of the gesture. In the multisensory and totalising atmospheres of hospitals, she argues, one sensory aspect gestures toward a whole atmospheric experience or gestalt. Drawing on gestalt theory, sensory ethnography and key texts on atmosphere she argues that the raised aesthetic awareness in the museum allows for the synaesthetic experience of atmospheric similarity across institutions.

Finally, in Chapter 12 Friedlind Riedl challenges current theoretical thinking on atmosphere and introduces the philosophical thought of German phenomenologist Hermann Schmitz. Through her ethnographic descriptions of communal singing in Southern Chinese music-making and feelings of shame and love, Riedl offers a critical re-reading of Böhme's conception of atmosphere. She proposes to move beyond thinking of atmospheres as topological and argues for a focus instead on duration and temporality. Rather than conceptualising atmospheres as constellations, she argues for a situation ontology based on observing the dynamics and performativity of everyday practices and events.

To conclude, this volume aims to contribute to the debate and incite more in-depth critical research into the role of atmosphere in everyday life. It is our contention that ethnography affords a particularly apposite means to investigate the atmospheric. It allows us to attune to the multifaceted layering and moods of everyday life and hence to investigate the thick fabric from which atmospheres are made and that in turn create the living beings whose lives they shape.

Bibliography

Anderson, B. 2009. Affective Atmospheres. *Emotion, Space and Society* 2, 77–81.

Anderson, B. and Ash, J. 2015. Atmospheric Methods. In *Nonrepresentational Methods: Re-Envisioning Research*, edited by P. Vannini. London: Routledge, 35–51.

Anderson, B. and Wylie, J. 2009. On Geography and Materiality. *Environment and Planning A* 41, 318–335.

Bateson, G. 1936. *Naven: A Survey of the Problems Suggested by a Composite Picture of the Culture of a New Guinea Tribe Drawn From Three Points of View*. Cambridge: Cambridge University Press.
Benedict, R. 1934. *Patterns of Culture*. Boston, MA: Houghton Mifflin Company.
Bille, M. 2015. Hazy Worlds: Atmospheric Ontologies in Denmark. *Anthropological Theory* 15(3), 257–274.
Bille, M., Bjerregaard, P. and Sørensen, T. F. 2015. Staging Atmospheres: Materiality, Culture, and the Texture of the in-Between. *Emotion, Space and Society* 15, 31–38.
Bille, M. and Sorensen, T. F. 2016. *Elements of Architecture: Assembling archaeology, atmosphere and the performance of building spaces*. London: Routledge.
Böhme, G. 1993. Atmosphere as the Fundamental Concept of a New Aesthetics. *Thesis Eleven* 36, 113–126.
Böhme, G. 1995. *Atmosphäre*. Frankfurt am Main: Suhrkamp Verlag.
Böhme, G. 2013. *Architektur und Atmosphäre*. Paderborn: Wilhelm Fink Verlag.
Debaene, V. 2014. *Far Afield: French Anthropology Between Science and Literature*. Chicago: The University of Chicago Press.
Durkheim, É. [1912] 2001. *The Elementary Forms of the Religious Life*. New York: Oxford University Press.
Durkheim, É. [1895] 2013. *The Rules of Sociological Method*. New York: Free Press.
Edensor, T. and Sumartojo, S. 2015. Designing Atmospheres: Introduction to Special Issue. *Visual Communication* 14(3), 251–265.
Geertz, C. 1973. *The Interpretation of Cultures*. New York: Basic Books.
Griffero, T. 2014. *Atmospheres: Aesthetics of Emotional Spaces*. Oxford: Routledge.
Harvey, P. 2013. Anthropological Approaches to Contemporary Material Worlds. In *The Oxford Handbook of the Archaeology of the Contemporary World*, edited by P. Graves-Brown, R. Harrison and A. Piccini. Oxford: Oxford University Press, 54–66.
Heidegger, M. [1927] 1996. *Being and Time*. Albany: State University of New York Press.
Ingold, T. 2007. Materials Against Materiality. *Archaeological Dialogues* 13(1), 1–16.
Ingold, T. 2011. *Being Alive: Essays on Movement, Knowledge and Description*. Abingdon: Routledge.
Ingold, T. 2012. The Atmosphere. *Chiasmi International* 14, 75–87.
Ingold, T. 2015. *The Life of Lines*. London: Routledge.
Massumi, B. 2002. *Parables for the Virtual: Movement, Affect, Sensation*. London: Duke University Press.
Mauss, M. [1925] 2002. *The Gift: The Form and Reason for Exchange in Archaic Societies*. London: Routledge.
Mauss, M. [1926] 2007. *Manual of Ethnography*. New York: Durkheim Press/Berghahn Books.
Merleau-Ponty, M. [1964] 1968. *The Visible and the Invisible*. Evanston: Northwestern University Press.
Morris, B. 2003. *Anthropological Studies of Religion: An Introductory Text*. Cambridge: Cambridge University Press.
Nuckolls, C. W. 1995. The Misplaced Legacy of Gregory Bateson: Toward a Cultural Dialectic of Knowledge and Desire. *Cultural Anthropology* 10(3), 367–394.
Orwell, G. 1946. *The Moon Under Water*. Available at http://theorwellprize.co.uk/george- orwell/by-orwell/essays-and-other-works/the- moon-under-water/ [Accessed 10 May 2016]
Pink, S. and Leder Mackley, K. 2016. Moving, Making and Atmosphere: Routines of Home as Sites for Mundane Improvisation. *Mobilities* 11(2), 171–187.

Riedel, F. 2015. Music as Atmosphere: Lines of Becoming in Congregational Worship. *Lebenswelt* 6, 80–111.

Schmitz, H. 2014. *Atmosphären*. Freiburg: Alber.

Schmitz, H., Mullan, R. O. and Slaby, J. 2011. Emotions Outside the Box – the New Phenomenology of Feeling and Corporeality. *Phenomenology and the Cognitive Science* 10, 241–259.

Stewart, K. 2011. Atmospheric Attunements. *Environment and Planning D: Society and Space* 29, 445–453.

Thibaud, J. P. 2015. The Backstage of Urban Ambiances: When Atmospheres Pervade Everyday Experience. *Emotion, Space and Society* 15, 39–46.

Throop, J. 2014. Moral Moods. *ETHOS* 42(1), 65–83.

Tsing, A. 2012. Unruly Edges: Mushrooms as Companion Species. *Environmental Humanities* 1, 141–154.

Tsing, A. 2013. More-Than-Human Sociality: A Call for a Critical Description. In *Anthropology and Nature*, edited by K. Hastrup. London: Routledge.

Zumthor, P. 2006. *Atmospheres: Architectural Environments, Surrounding Objects*. Basel: Birkhäuser.

2 Hauptschule

Atmospheres of boredom and ruination

Stefan Wellgraf

Introduction

Atmospheres of boredom are experienced individually. At the same time they are observable beyond specific actors and interpretations. At the school in Berlin where I was doing ethnographic research, the term "boring" was among students' favourite adjectives. For example, students commenting on the at times deviant behaviours of their peers – ranging from goofing around to randomly destroying school property – typically observed that those students "must've been bored". Occasionally, they added "very" in an attempt to raise the bar of "bored" in their overall negative perception of the school around them. My goal in this chapter is to show how my field observations on boredom can be systematised, and linked to discussions on atmospheres and infrastructural ruination in order to develop a political reading of the socio-cultural production of affective atmospheres in Berlin Neukölln's schools for the poor.[1]

The notion of boredom is difficult to grasp but is nonetheless an ubiquitous phenomenon. It is at once "an experience without qualities" (Goodstein 2005) defying description and a "delicate monster" with a voracious and indiscriminate appetite. Although the articulation of boredom at the Galilei school is frequent, "boring" is attributed to a variety of distinctive situations and conditions. The term's inflated usage points to the limits of our verbal expressiveness. As a categorisation derived from the field, boredom designates more than the description of conditions or circumstances and becomes an abbreviation or slogan for a general uneasiness of students and teachers at the school. The exclamation "I am bored!" in the classroom is an avowal of feeling as well as an accusation. Even though the students articulate a subjective state, their feeling conveys the normative claim that school should really be more interesting. Hence expressions of boredom are also a situated critique of an education perceived as inadequate and unsatisfying.

In this chapter I understand boredom neither as a classical social emotion or personal feeling like love, anger or jealousy, nor as an affect in the sense of a pre-personal or pre-social experience that cannot yet be classified (Massumi 2002). Rather, going along with Ben Anderson's (2009) reading of affects I understand boredom in Berlin's Hauptschulen as 'affective atmospheres', that are strongly felt by individuals yet are perceptible beyond the boundaries of the subject. I will use phenomenology (Böhme 1995), but also go beyond it by understanding the affective atmospheres not as something universal, coming from nowhere, but as situational and contextual. The main aim of the chapter will be to show that the atmosphere of boredom at the Galilei school emerges as a multi-layered phenomenon that is dependent upon a particular historical context as well as the material, social and cultural dimensions of life at the school. I will begin my discussion with a brief categorical reflection introducing a differentiation between atmospheres, feelings and moods. Highlighting historical processes, I will then proceed by bringing the affective atmosphere of boredom into discussion with the notion of ruination. Next, I will move to an ethnographic description of boredom in the school, distinguishing temporal and spatial dimensions of boredom in order to highlight its situatedness within a nexus of material and social practices. My overall aim is to provide both empirical and theoretical suggestions of how the ethnography of atmosphere can become a tool for political critique.

Atmospheres, moods, feelings

Before entering the school, some brief reflections on what I mean with the seemingly fuzzy description "atmospheres of boredom" seem to be necessary. Clarifying first the understanding of boredom and of atmosphere and then differentiating between atmospheres, feelings and moods is crucial as it lays out some fundamental methodological and theoretical questions in the current debates on atmospheres, which will later be further elaborated in the empirical sections.

The perception of situations as boring emerges in the context of modernisation concomitantly with the assertion that time is abstract (Goodstein 2005; Meyer Spacks 1995). The English word "boredom" and the German "Langeweile" come into usage in the 18th century as distinctive from older, related terms like "acedia," "melancholy" and "ennui". Not going into an in-depth discussion of historical semantics, I want to develop at least some provisional terminological distinctions concerning the analytic register of boredom. I understand the situational and existential experiences of boredom as affective atmosphere (Bondi et al. 2007; Anderson 2009; Lehnert 2011). By linking atmospheres to the affective turn in the social sciences and humanities Anderson's approach serves us in two ways. Firstly, he points out that atmospheres are ephemeral and shifting phenomena that at the same time are important aspects of the affective qualities of everyday life. Secondly, he underlines that the very ambiguity of atmosphere unsettles any rigid distinction between affect and emotion, such as linking affect with the impersonal and objective and emotion with the personal and subjective. Following this incitement, I understand the atmosphere of boredom as always contextual rather than as an objective quality. Atmospheres designate the aesthetic-affective qualities of spatio-temporal environs and situations. Anderson refers here to new German phenomenology, which in a similar way tries to conceptualise these kinds of qualities by understanding the concept of atmosphere as an emotional, bodily and sensual experience (Böhme 1995, 2001, 2006; Schmitz 2007). The term atmosphere has been elaborated in this context as a "funcamental concept of a new aesthetics" (Böhme 1995), an aesthetics not only limited to the realm of the arts but also encompassing everyday practices and experiences. However, the weakness of the phenomenology of atmosphere seems to be its lack in addressing the role of race, class, gender and other powerful forms of inequality in their social production, something that I attempt in the chapter by looking at a lower-class school populated mainly by migrant students.

Given the wide range of conceptions of atmosphere used in both everyday speech and academic discourse, I do not intend to fix this ambiguous term with a strict definition, but want to point out here some possible demarcations from sister terms such as mood (or Stimmung in German), taking German philosopher Thomas Fuchs (2013) as my guide. According to Fuchs, moods describe an affective colouring of life or episodes in life that go beyond specific spatial arrangements and are accompanied by corresponding perceptions, judgments and behaviours (Wellbery 2003; Gisbertz 2011; Gumbrecht 2011; Reents and Meyer-Sieckendick 2013). Put differently, moods are difficult to delineate and to situate spatially because they are not contained by particular situations but rather shape the ways in which we perceive these situations. Feelings, on the other hand, Fuchs suggests, are more short-lived than moods and tend to take center stage. Even though they have a higher felt intensity, they may quickly dissolve. Atmospheres, on the other hand, have both a spatial and a temporal dimension, both elaborated in more detail below. They have a strong affinity with moods, but at times can also be experienced as forceful feelings that take centre stage.

What can be gleaned from Fuchs characterisation, atmospheres, moods and feelings tend towards a certain conceptual overlap and empirical attunement. In the context of my ethnographic material this means that a very basic boring mood can produce in some people at certain moments an unbearable feeling of boredom, and the repeated onset of such situations perceived as boring can lead to a permanent, all-encompassing atmosphere of boredom. In other words, at the Galilei school, boredom is both at the foreground and at the background. As a pressuring feeling, boredom affects space-time situations that those involved would much rather escape; it is reminiscent of situational boredom. As latent mood, it frames experience, and sheds a certain light on what is happening. Among the peculiarities of atmospheres of boredom is its capacity for making spaces grey and wan and time dull and vacuous (Svendson 2002). However, despite boredom's reign in the Galilei school, disparate spatial elements, dissonant moods and emotional counter-reactions like being in love, joking around or "trash talk" can also be observed.

Bringing my argument to the fore that atmospheres are contextual, that they are socially and materially produced and not given, I will now explore the current school's boredom in a historical and social context. Here the concept of ruination with its focus on infrastructure will serve me to bring out the processual and relational character in the making of atmospheres.

Ruination: the decline of a school

The spatial infrastructure of the Galilei school grounds is systematically being ruined. Infrastructure is generally associated with material supply systems, like networks of roads and public transportation, but school infrastructures can also be understood as facilitating conditions for learning and educational advancement

(Star 1999; Larkin 2013; Angelo and Hentschel 2015). Infrastructures influence subjectivation processes by positioning the students in social space and by forming practical routines and attitudes. The dimension of power in infrastructures reveals itself through a relational perspective; for example, through a comparison between material facilities in different kinds of schools (Schmidt 2002; Allmendinger and Leibfried 2003: 73). At the same time, actors get an idea of the larger social correlations by observing the conditions of infrastructures. The Galilei school's spatial infrastructure meets aesthetic and operational minimum requirements only in part. The spatial basis for instruction is provided, but beyond that, during recess, for example, the school's resources are insufficient. As a gradual process of disintegration, deficient infrastructures often become visible in moments of disruptions and crisis (Graham 2010), as has been the case in Berlin, when abuses at individual schools are discovered by chance. What remains usually hidden are invisible ways of withholding sufficient resources for maintenance. Casual comments about the rarely seen cleaning personnel or the absent gardener indicate how the once splendid school building has been run down over time.

In the context of postcolonial infrastructures, the US historian Ann Stoler has argued that their ruination is controlled by politics through spatial processes of exclusion (Stoler 2008, 2013; Street 2012; Larkin 2013; Martin 2014). These processes can certainly be confirmed in Berlin, where the complaints about the disadvantages of Berlin's district Neukölln, described above, suggest that the resources for the renewal and maintenance of schools are unequally distributed. From a postcolonial perspective that sees racist exclusions as happening in the present, the increasing spatial negligence of the Galilei school since the 1970s and its increasing immigrant student body are expressions of the racial hierarchisation of the German educational system. Berlin's affluent Zehlendorf, on the other hand, symbolises not only wealth but also 'Germanness', whereas Neukölln comes to stand for poverty and a foreign student body. While the Galilei school is not in ruins, I suggest that the subtle ways of its gradual ruination allow for a political reading of the socio-spatial processes of becoming a ruin.

Attending to processes of ruination can illuminate the historicity of atmospheres and how they develop over time. School buildings and other infrastructures of modernity carry affective traces of their past into the present time (Navaro-Yashin 2009; Larkin 2013). At the Galilei school, the splendour of the past can be gleaned in places like the imposing auditorium, big enough for the entire student body. The former splendour adds even more to the present impression of decay. The Galilei school building opened its doors with much fanfare on 19 May 1929. The new school was minted in the spirit of the Weimar Republic, departing from the typical Prussian military architecture and distinguishing itself also from radical reform ideas. Most school buildings at the time harked back to the latter part of the 19th century, the so-called *Gründerzeit*, when schools with long hallways took their spatial inspiration from military barracks and monasteries (Göhlich 1993; Schneider 1998). The new school building had classrooms along a central corridor, but with its large windows and wide stairwell was also conceived under the bright and generous influence of 'New Objectivity', with its functional approach to architecture.

Numerous rooms dedicated to manual tasks, art and athletic activities introduced another innovation. They included rooms for crafts and experiments; two darkrooms sponsored by the Agfa Company; broadcasting equipment for teaching foreign languages; for sports, a hall with showers and toilets akin to those in "sophisticated hotels" (Homann 2001: 138), a main auditorium for six hundred people, and lastly, an observatory under the rooftop, as special feature. The interior of the Galilei school's classrooms marked a compromise between the reform-pedagogy of the 1920s and the former status quo: instead of sitting on benches as they had previously done, students now sat in twos at tables that, as before, were directed to the front of the room. Students facing the front remained a principal arrangement in the schools of the Federal Republic of Germany after 1945, and it is still existent today at the Galilei school.

The historical turbulences of the 20th century in the period following had a strong impact on the school. A majority of the school's faculty welcomed the National Socialist rulers and most of the students belonged to Nazi youth groups. During the war, the school was turned into a military hospital, the classes moved elsewhere, and many graduated early only to get drafted immediately afterwards or to work in the *Reichsarbeitsdienst*, the official state labour service of Nazi-Germany. More than a third of the student body perished in the war. The school building, however, was hardly damaged and served its purpose once again in 1946. The main auditorium was renovated in the early 1950s and, as one of the few unimpaired places fit for meetings, became a cultural centre of post-war Neukölln. The Cold War period affected the school profoundly, due to its close proximity to the Wall. As part of the partitioning of Berlin and the construction of the Wall in the early 1960s, 100 students were barred from reaching the school; 25 of them soon escaped to West Berlin.

In the decades to follow, the school became increasingly a domicile for outsiders and migrants. With that came social tensions and structural disintegration. After several name changes and reorientations, the last graduates left the venerable, old *Gymnasium*, traditionally the highest and most respected institution of secondary education in Germany, in 1963. It was irretrievably dissolved and the new school, then called Galilei school, was founded. Now a *Hauptschule*, symbolising originally the main type of secondary education but soon being regarded as its lowest and least respected form, it became characterised by conflicts between Turkish and German students, with the latter staging a school strike against the newcomers. When the Berlin School Board separated students with a migrant background according to nationality, it allocated to the Galilei school the remaining 'rest of the world'. Even today, Galilei school students come from many different countries, and many of them from Palestinian, Kurdish and Yugoslavian refugee families. When in 2010 new structural changes occurred, that led to the merging of *Hauptschulen* and *Realschulen* into new models for secondary schools; the Galilei school had no *Realschule* to merge with. During my fieldwork, I attended tenth grade classes that comprised the last graduates of the now obsolete *Hauptschule*. At the time of my research in 2012–2013, the number of students from a non-German language of origin was at 88.4 percent. Ninety-two percent of the students were exempt from having to contribute financially to the purchase of learning materials. In other words, I encountered a predominantly immigrant student body whose families survived almost without exception with state transfer benefits.

A cursory overview of the spatial and institutional history of the current Galilei school suggests a marked decline. One of the most "progressive" schools in one of Germany's most "beautiful" and "modern" school buildings turned into a school often pejoratively called a "problem school", "high risk school" and "hotbed of violence" by the Berlin press. This history of decline is also a history of spatial-temporal changes – from refined to almost unusable school toilets, from the initial pride of Neukölln students in their school to their shame and boredom. The critical point here is not that students are occasionally unsatisfied by certain teachers or individual subjects – something that occurs at all schools – but that they are on the whole disappointed by their school. Their discontent is voiced in the high rate of absences and a widespread scepticism towards all of the school's offerings, a scepticism that is shared at least initially even by those who try to make the offerings attractive to the students. As a result, distance among students and scepticism among teachers creates an atmosphere of boredom and alienation.

The decline of the Galilei school shows itself not only in an historical process of degradation, the ruination continues today and affects both the physical condition of spatial infrastructure as well as the negative perception of school-space and school-time by current students and teachers. How this leads and further contributes to an atmosphere of boredom will be shown by looking ethnographically at the spatio-temporal dimensions of boredom at the school.

Spaces of boredom

The literature on boredom commonly differentiates between situational and existential boredom, with the former characterised by external, short-term circumstances and the latter by long-term questions about meaning. In the context of education, this differentiation empirically involves the spatial perceptions and temporal horizons of the quotidian life at the school. Tracing the atmosphere of boredom, then, necessitates the conceptualisation of time and space as socially constructed. Following Henri Lefebvre (2004), I understand space neither as a neutral, physical container nor as a purely subjective idea. Instead, I pursue it in its production as a set of appropriation practices that are available or inaccessible. That way, buildings for instance appear less as static, permanently fixed structures and more as continuously changing constructs whose condition and meaning changes – in our case, into a slowly ruined space (Gieryn 2002; Delitz 2009). It follows that school architecture is a product of society and concomitantly, a social medium. Analogously, I take time as historically mutable, culturally specific and even within a given society, distinctly marked by milieu. Situational and existential forms as well as spatial and temporal dimensions of boredom are interrelated. Keeping both sides of the coin in mind is requisite to comprehending the complex nature of boredom. While quantitative-sociological methodologies cannot grasp the depth of boredom, humanities-based approaches tend to miss out on

Figure 2.1 Empty school-yard
Photograph by the author

its profane forms and social conditions. My own ethnographic approach focuses on concrete situations of boredom in the daily life of a school and discovered in seemingly banal moments spatial and temporal textures as well as recurring patterns of overarching problems and situational reactions.

The layout of the campus illustrates the spatial component of the school's atmospheric misery. With no one to take care of the yard, nature is reclaiming the space. Trees and bushes are discreetly encroaching, unwittingly providing smoking students some protection from the scrutinising eyes of their teachers. In the midst of this grassy landscape, yellow dandelions grow amongst weeds in the summer and dirty snow or slippery ice cover the surface in the winter, with no one removing either. A basketball hoop on the edge of the schoolyard and a rusty soccer goal are the only recreational offerings. When I talked to Holger Thomalla, the janitor, about the schoolyard, he seemed frustrated: "We've been wanting to do something about it for years, but they always say there's no money. In Zehlendorf, a schoolyard makes a different impression. But nobody's interested in Neukölln, and it shows". His comments regard the social environment of Berlin in terms of a hierarchy defined by spatial oppositions, a mechanism that Pierre Bourdieu (1997) teased out in connection to the disadvantaged infrastructures of the Parisian banlieues. Social environments are characterised by mutual exclusion, whereby the relational positioning in an environment indicates the social position within the larger power structure. Wolfgang Rüttgen, the school principal, also drew comparisons to the more affluent district Berlin Zehlendorf and saw the reasons for the desolate condition of the school in underfunding and the lack of parents' involvement. In addition to the inequality between poorer and more affluent districts, the inequality within Neukölln was also pointed out. While the school campus Rütli, which has become a showcase for urban educational rehabilitation, received millions, the rest of the schools in the area went empty-handed. In addition to cuts in the maintenance of the schoolyard, a shortage of money for cleaning the school building has created a massive garbage and cleanliness problem. Hallway floors were rarely scoured; toilets became filthy; and garbage was littering the schoolyard.

The school's institutional decline in the past few decades articulates itself at present in an atmosphere of boredom. To link the historical process of ruination to the current discussion of atmosphere it might be helpful to turn for a moment to German phenomenology. The philosopher Hermann Schmitz, whose impact has often been neglected in English-speaking academic discourse on atmosphere, turned to the realms of the lived-in world. Schmitz (2007) conception of feelings is closely related to his understanding of corporeality and space. He conceives

Figure 2.2 Nature is reclaiming the school space

Photograph by the author

of the body as sensing and of feelings as bodily affliction (ibid. 23). For Schmitz feelings are atmospheres, but feelings no longer bound to the notion of an interior subject. In recent years, the philosopher Gernot Böhme (1995, 2001, 2006) has contributed to the idea of atmospheres by extending it first to the realm of art and aesthetics, then to the realm of the quotidian, and finally, to the basic principle of entire theory of perception. Böhme points out that atmospheres are primarily perceived spatially and emanate from objects within them, which he refers to as "attuned spaces" (Böhme 1995: 45). He therefore understands atmospheres as being constituted in the in-between of perceiver and perceived and as arising from a particular spatial constellation of things in an environment. Both Schmitz and Böhme, however, tend to attribute an objective character to feelings and atmospheres, as if those existed independent of social actors. As basis for an analysis of the production of space, I argue, both philosophers' phenomenological approach must in practice and theory be situated temporally and locally: spatial feelings happen in a place and at a time; they are, moreover, culturally coded and thereby entangled in the habits and routines of agents (Reckwitz 2012). Such processes are connected to spatial arrangements and practices in an atmosphere of decay. Indeed, their interrogation opens up the historical, cultural and social dimensions in the production of affective space such as atmospheres of boredom at the school.

Despite their tendency to objectify atmospheres there are observations and proposals in both Böhme's and Schmitz's writings which can help our understanding of boredom in the Galilei school and of atmospheres more generally. Böhme mentions, for example, the aesthetic right to live in an environment that one is allowed to help create and in which a feeling of well-being is conceivable (Böhme 1995: 40). Such basic aesthetic needs are hardly met by a rundown schoolyard. Students are unmoved by the situation at the Galilei school, or so they say. They take it in stride and talk about spatial conditions and creative possibilities with a hefty dose of distance and alienation.

The spatial production of an atmosphere of boredom can be followed also inside the school-buildings. During my fieldwork I often observed the typical bodily postures of the students that were reminding me of the formal illustration of an inventory of the gestures that Peter Toohey (2011) has observed to be signs of boredom: folded arms, elbows on the desk supporting a weary head, eyes staring into the void. A wide disinterest and shared boredom in the classroom was typical for lessons and rarely changed. In order to explore the spatial creation of boredom, I would like to focus here more explicitly on the room's spatial dimension.

The classroom follows an ordering principle that was considered progressive when the school opened in the late 1920s, gained acceptance in the post-war period, and replaced the school benches arranged in rows once and for all: two students are sitting next to each other at a table facing the teacher, in hierarchical fashion (Müller 1998). The skid chairs and skid tables at the school were patented in 1950 (ibid.). Back then the model was considered to be easy on people's backs; it did not damage the floor, and was easy to lift for those cleaning the classrooms. After 1973, the skid chair was made cost effective with oval steel tube legs and plywood seats. This model remains popular despite the fact that flexible chair backs are considered more back-friendly today. But is this ordinary school furniture suggestive or creative of a boring atmosphere? I would suggest it is on closer inspection of its uses and arrangement. Students are hanging in their chairs and resting on their tables. Unoccupied chairs remain standing on the tables and were not, as teachers sometimes demand, lowered to the floor. These observations indicate the distance between students and their classroom, which does not become a space in which they feel comfortable or intellectually challenged.

The way in which specific school environments shape relations between people and artefacts can be usefully comprehended through spatio-sociological approaches. Martina Löw (2001), a German sociologist, conceives of space as a relational configuration of artefacts and people in places. This configuration is created in and through positioning, alliances and synthesis – as a set of activities rather than as an a priori organisation based on architectural structure. Education researcher George Breidenstein (2006) has developed Löw's insights by showing through his ethnographic material how classrooms emerge from overlapping visual, acoustic and haptic spaces. Breidenstein's differentiation is useful to a more thorough understanding of the production of space at the Galilei school; consider a description of my video recordings:

> Mr. Stein is standing at the window in the front of the classroom and explains a math problem to two girls. In the middle of the room a few boys are congregating around Jamal's table, chatting. Two of them walk towards the teacher and then, standing right next to him, they grope one of the girls' note books, briefly look around, and return. Another boy does push-ups in the isle between the tables. He's still wearing his headphones in one ear, after listening to music during class. After ten perfect push-ups, he is joined by another student. They complete another ten push-ups. Intermittently, they check their form – motivated by the remarks of their animated peers. One student jokingly sits with half of his body weight on one of the boys exercising. Afterwards they all assess the scene with expert lingo. Two, three minutes have passed, and the teacher is still busy with the girls in the front. When he is finished, he looks up and advises students to take their seats. Most of them saunter back to their chairs and continue their conversations in small circles.

The above scene from a 10th-grade math class appears like a state of anomy: rules are powerless; the organisation is atrophied; and digressions are the norm. The educational space created in the situation is momentarily defined by a partial destructuring and a new hierarchisation. While the scene unfolds in a typical classroom, the spatial elements have been reappropriated, and the teacher intermittently lost control to a group of tenth graders – on all levels: visual, acoustic and haptic. Visually, the teacher is restricted by his focus on the two students sitting directly in front of him losing sight of everything else. Even though he manages to get them to return to their seats, his loss of authority is apparent in the way students simply ignored him in close proximity. The teacher barely even tries to regain his authority. While two boys and a group of girls are sitting quietly doing their work, for a few minutes a handful of students take over the sovereignty of the classroom by wandering around, forming groups, and misusing the classroom for their athletic activities. Their territorial claim is accompanied by an acoustic dominance. Whereas they previously distracted themselves listening to music and occasionally heckling, now their conversation takes over the room, while the teacher standing off-side is hardly audible. The students have also expanded their haptic space. They wander around with objects, touch in provocative ways the note books of the girls and use the floor for their strength exercises. Their message is clear: they feel neither engaged by the school nor by the teacher. Breidenstein (ibid.: 85) observed a silent consensus that tolerates occasional boredom and students' ways of making time pass as long as this is done quietly and the purpose of teaching left unquestioned. However, students at the Galilei school seem to denounce an implicit consensus. They do not want to put up with their boredom at school and provoke the teacher with their actions. In such cases, the whole school succumbs to an atmosphere of boredom.

Empty time: no meaning, no future

In addition to spatial dimensions, this atmosphere of boredom in schools has temporal dimensions that are crucial to consider if we want to understand their particular constitution. The contingency of time regimes, mostly taken for granted in everyday life, often only becomes visible as a result of conflict. The dominant reliance on time as an abstract emerged in the face of great resistance in the course of a modernisation that favoured a task-oriented structured workday and a content-oriented structured education (Urry 2000). The temporal structure of the school day is therefore the result of internalising and learning-processes that do not always run smoothly. Time regimes are not only historically founded, culturally informed and socially representative but also show variance within a society (Schell 2000). At the Galilei school, the time regime is continuously implored precisely because the school has reached a crisis in meaning and legitimacy. Consider this excerpt from my fieldnotes:

> Because of the sludge, my bicycle stays at home in the morning and I stagger half asleep to the bus station. The bus is late. Many students depend on the bus and therefore a whole bunch of them will arrive at school about fifteen minutes late. Only a few of the young ones attempt to hurry. The entrance to the school is locked. The students look at me inquiringly, but I do not have a key. It is mighty cold and we are standing right below the window of the teachers' lounge. The ethics teacher shows some kindness and secretly unlocks the door about five minutes later. The doors of the classrooms are also usually locked by this time, but some teachers let the students enter regardless; the other students wander around like ghosts in the school building. In his lecture on punctuality, the principal later refers to the events of the morning and rages at the students: "Employers aren't interested in buses running late. And those who let the students in are uncooperative – both towards teachers and students. If you cannot learn to be punctual, you are not fit for life."

With his admonition, the principal falls back on the primary social function of educational institutions to teach socially dominant time regimes and thus the transition into work life. The associated discipline usually follows from the internalisation of temporal guidelines that are taken for granted: students are expected to show up at 8:00 in the morning, just as they will later have to pursue their work according to a similar rhythm. The social guidelines can only be put into practice by means of temporal measuring and ordering devices. Time of day and precursors of daily class schedules first appear in German school regulations in the 15th century (Luhmann 1975, 1990; Dohrn-van Rossum 1995; Macho and Kassung 2013). The meanwhile established school clocks and class schedules make time concrete and a time regime enforceable, which in turn connects the school to the larger social order. Historically, the school has thus contributed substantially to the hegemony of a linear and abstract conception of time, first in Western Europe and later, in other parts of the world. Time became an instrument of power and schools and clocks, symbols of modernisation.

Today at the Galilei school, however, time is out of joint. A de-structuring of the time regime is threatening the school's functionality. Locked entrances and the indignation of the principal are reactions to a problem whose causes point far beyond the school and whose consequences are nonetheless impacting the entire daily life at the school. With the failure to provide students with the path to a working life, the school's time regime is in danger of losing its horizon (Drews 2008: 24–25). School education loses its long-term purpose and disciplinary actions based on a timeline largely forfeit their legitimacy. The decay of time at the Galilei school is imminent. Obvious indications are the heavy rate of absences and tardiness; some students are gone for days or weeks, while others only attend a couple of hours in the afternoon to socialise with their friends.

The hollowness of educational attempts and the student's bleak perspectives on the job-market contribute to a perception of school-time as empty and meaningless. This can be seen as a contemporary example of alienation. The term alienation refers historically to a set of experiences similar to those of modern boredom, yet carries different consequences. Since the 19th century, diagnoses of alienation have served to convey uneasiness with the modern, but their theoretical context is Marxist and their goal, social change. Marx's (1973) notion of alienation was a critical response to the troubled relations between workers and the process and product of their labor, brought about by industrialisation and mechanisation. Linking ruination up with alienation might bring us a step further in assessing the political dimension of educational atmospheres of boredom.

Contemporary Frankfurt School theorists like Rahel Jaeggi (2005) and Hartmut Rosa (2009, 2012, 2013) analyse today's forms of alienation from the material and social world by regarding relations to self and world as troubled due to existing social structures and institutions. In her work on alienation, Jaeggi formulates it as a "relation of unrelatedness" (Jaeggi 2005: 19). To her, alienation is not simply the absence of a relation but rather a deficient connection, which in our case would point to the missing positive identification with the school. The noncompliance with time schedules, the classes that are perceived as senseless, and the ensuing boredom exemplify processes of alienation from an institution that is so disappointing precisely because students cannot be impervious to it. Hartmut Rosa (2009), too, considers the notion of alienation as a key category for social criticism. In this context, his reflections on the idea of resonance offer insights into the topic school and boredom (2012). The drab schoolyard and the empty school years inhibit students from experiencing a positive resonance with a school that is merely dull and dry.

There are further indications of the chipping away of the temporal order in the Galilei school that are indicative of a sense of alienation. Upon entering the school building, students immediately search out the teacher substitution schedule, which is obscured by the many students trying to catch a glimpse of it. Due to the teachers' frequent sick days, hardly a day goes by without classes being cancelled or teachers being substituted. As a consequence, certain subjects are not taught for an entire six months and school days on which solely substitute teachers helm the classes are not a rare occurrence. At other times the students are completely without supervision. The separation between lessons and recess seems tenuous. The students frequently ignore the bell, arrive late to class, or simply leave during class. Although the students only vaguely orientate themselves to the ring of the school bell – an electronic buzzing sound at the Galilei school – the teachers desperately harp on it, as if to demand the return to an operational order at school. Because of the chipping away at class time, a specific rhythm emerges, a spatial-temporal pattern of school interaction (Lefèbvre 1992). This rhythm is less determined by the switch from tension to release than by phases of passivity that are disrupted at moments with aggressive heckling or silliness. Locked doors are a desperate measure to save time from fully disintegrating, but a measure that is only partly enforced and shows unpleasant side-effects, like students walking around on the school grounds at will. The designated class time is after all without content and effective teaching is hardly possible as this section from my field diary highlights:

> When the door bell rings – more like the sound of electronic buzzing – the remaining handful of students who came late to school are now allowed to enter the classroom for the rest of the 90 minute class in occupational orientation. Without excuse, they trot to their seats. Instead of working, like the other students, on their wall journals, they do nothing. It doesn't seem worthwhile in their minds, since the work on the wall journals at any rate will continue for several more weeks. The mood is subdued and weary, typical for a Monday morning. One of the guys yawns and puts his head between his arms, one of the girls stares at her painted fingernails for a long time. "I am bored," yells one of the students after a while, but he gets no reaction. A student gets up, takes the clock above the door off the wall and starts to repair it. Watching the scene, the teacher remarks that it is high time a new clock was seen to.

The broken clock is a convenient symbol for time standing still in a state of boredom and a metaphor for the disjointed connections between the historical trajectory of education, the educational present day, and the professional future.

Time, according to the system of education, looks to the future; the purpose of learning unfolds before a horizon of opportunities. The missing prospects for the future among the Galilei school students unsettle their self-perception. Schools are called on to do their part in enforcing the discipline of time, but the teachers are lacking convincing, long-term incentives for such an enforcement. Since over 90 percent of the students come from families dependent on state assistance, many of them are raised in homes impacted by unemployment and have little experience with the model and timeline of a typical workday. A sprawling atmosphere of boredom is the consequence of school time's disintegration.

Lamentation about boredom in the scene described above arises from a particular situation: it is Monday morning and the grey weather as well as early morning tiredness do their part in making everyone passive and time never-ending. That day I am among those fighting sleepiness. My tired eyes prompt students to ask me whether I did drugs on the weekend, and later, the teachers offer me coffee. But the specific situation represents at the same time the larger problem of existential meaninglessness and futurelessness that is especially virulent when it comes to the subject of occupational orientation. The larger problem leads to a depressing mood that cannot be limited to bleary Monday mornings. Such moods often reveal themselves in passing details or irritating moments (Gumbrecht 2011: 29), for example, in the chairs that remain standing on the desks, the students who are late and in no hurry, or the broken school clock.

Conclusion

Classical and contemporary diagnoses of boredom in modernity not only took stock but always also offered critiques. Martin Heidegger's ([1929/30] 1983) emphasis on "deep boredom" (ibid.: 111) encapsulates an existential type of intellectual being-left-empty that offers a deeper view into Being. Such a reflection rooted in boredom may lead at the Galilei school to questions like "What am I doing here?" or "What's it all about anyway?". Furthermore, other critical theorists went beyond seeing boredom as some neutral diagnosis of time in slow motion and saw political potential in the uneasiness of perceiving time as empty and atmospheres as boring. Siegfried Kracauer (1977) called deep boredom "radical boredom" (ibid. 321–325) because it may go hand in hand with critical reflections on the circumstances of one's own life. Walter Benjamin, who looked for illuminations even in the profane, imagined boredom as a kind of continuous, languorous yet accentless slumber – an awakening would be the liberation from the nightmare of history (Benjamin 1982, 2007). Furthermore, Henri Lefèbvre discovered in boredom the traces of unrealised desires and opportunities and therefore assigns to it a utopian potential (Lefèbvre 1974; Gardiner 2012). In the light of these views, I suggest, that in the many complaints about boredom at the Galilei school resonates a form of situated social critique, expressed through vocabulary of personal dissatisfaction. Atmospheres of Boredom, then, may create room for self-reflection that opens up the potential for existential questions about the meaning of schooling to emerge.

However, boredom at school and the affiliated diversionary tactics are rarely perceived as critical reflection and more often as an individual refusal to learn or as a disruption of class. Boredom is frequently brought in relation with impulsive behaviour and is often treated as a symptom of attention deficit disorder (Boden 2009; Pease 2012). Privatisation, pathologising and medicalising of boredom contribute substantially to the general disregard for the social causes and dimensions of a subjective experience and to the divestment of its critical potential.

Building on my field research conducted at a secondary school in Berlin, I reconstructed the social production of different dimensions of boredom and linked them to discussions on atmospheres and ruination. Seeing the present-day affective atmosphere of boredom inside the school connected to processes of ruination enabled me to take both its spatial *and* temporal dimension as well as its historical *and* social production into account. In combining approaches stemming from different theoretical traditions, I showed that a thick ethnography of affective atmospheres needs to connect to different research-angles: subject- *and* object-oriented perspectives, social science *and* humanities approaches, narrative *and* material analysis, phenomenological *and* political reflections (Navaro-Yashin 2009). With an interpretation of atmospheres of boredom as results of ruination, I am offering a political reading that brings critical theory to the affective spaces of Berlin schools for the poor. Atmospheres of boredom in this context emerge not as simply given phenomenon, but as politically construed expressions of social inequalities. They are not only an isolated problem of today's students and teachers but are an effect of exclusionary processes and a fundamental problem of German society.

Note

1 The German secondary school system is divided into three layers. Today, polemically speaking, one might generalise that the German *Gymnasium* leads to university, the Realschule to working lives, and the Hauptschule to unemployment. The learning conditions at Hauptschulen have recently become problematic, and its reputation has deteriorated. They are typically attended by working-class students from a largely unemployed immigrant background. In 2008–09 I did ethnographic fieldwork in three Hauptschulen in different parts of Berlin (Wellgraf 2012); the study reported on here took place in 2012–13 at the Galilei school in Berlin-Neukölln. All proper names in this essay are fictitious in order to ensure anonymity. As a result of Berlin's restructuring of secondary education in 2010, the Galilei school was in the middle of transitioning from a Hauptschule to a Sekundarschule, the new designation for the fusion of Real- and Hauptschulen, at the time of my research. As participant observer I became involved in the daily life of the school for one year and made use of different research methods; in addition to observations and interviews, I utilised photographs, audio and film recordings.

Bibliography

Allmendinger, J. and Leibfried, S. 2003. Education and the Welfare State: The Four Worlds of Competence Production. *Journal of European Social Policy* 13(1), 63–81.

Andermann, K. and Eberlein, U. (eds.) 2011. *Gefühle als Atmosphären: Neue Phänomenologie und philosophische Emotionstheorie*. Berlin: Akademie.

Anderson, B. 2009. Affective Atmospheres. *Emotion, Space and Society* 2, 77–81.

Angelo, H. and Hentschel, C. 2015. Interactions With Infrastructure as Windows Into Social Worlds: A Method for Critical Urban Studies. *City* 19(2–3), 306–312.

Benjamin, W. 1982. *Das Passagenwerk*, Vol. I. Frankfurt/Main: Suhrkamp.

Benjamin, W. 2007. Über den Begriff der Geschichte. In *Erzählen. Schriften zur Theorie der Narration und zur literarischen Prosa*, edited by W. Benjamin. Frankfurt/Main: Suhrkamp, 129–140.

Boden, J. 2009. The Devil Inside: Boredom Proneness and Impulsive Behaviour. In *Essays on Boredom and Modernity*, edited by B. Dale Pezze and C. Salzani. Amsterdam and New York: Rodopi, 203–226.

Böhme, G. 1995. *Atmosphäre*. Frankfurt/Main: Suhrkamp.

Böhme, G. 2001. *Aisthetik: Vorlesungen über Ästhetik als allgemeine Wahrnehmungslehre*. München: Fink.

Böhme, G. 2006. *Architektur und Atmosphäre*. München: Fink.

Bondi, L., Davidson, J. and Smith, M. (eds.) 2007. *Emotional Geographies*. Hampshire: Ashgate.

Bourdieu, P. 1997. Ortseffekte. In *Das Elend der Welt*, edited by Pierre Bourdieu et al. Konstanz: UVK, 159–167.

Breidenstein, G. 2006. *Teilnahme am Unterricht*. Wiesbaden: VS.

Delitz, H. 2009. *Architektursoziologie*. Bielefeld: Transcript.

Dohrn-van Rossum, G. 1995. *Die Geschichte der Stunde: Uhren und moderne Zeitordnungen*. München: dtv.

Drews, U. 2008. *Zeit in Schule und Unterricht*. Weinheim: Beltz.

Fuchs, T. 2013. Zur Phänomenologie der Stimmungen. In *Stimmung und Methode*, edited by F. Reents and B. Meyer-Sickendiek. Tübingen: Mohr Siebeck, 17–31.

Gardiner, M. 2012. Henri Lefebvre and the 'Sociology of Boredom'. *Theory, Culture & Society* 29(2), 37–62.

Gieryn, T. 2002. What Buildings Do. *Theory and Society* 31(1), 35–74.

Gisbertz, A. (ed.) 2011. *Stimmung: Zur Wiederkehr einer ästhetischen Kategorie*. München: Fink.

Göhlich, M. 1993. *Die pädagogische Umgebung: Eine Geschichte des Schulraums seit dem Mittelalter*. Weinheim: Deutscher Studien Verlag.

Goodstein, E. 2005. *Experience Without Qualities: Boredom and Modernity*. Stanford, CA: Stanford University Press.

Graham, S. 2010. *Disrupted Cities: When Infrastructure Fails*. New York: Routledge.

Gumbrecht, H. 2011. *Stimmungen lesen: Über eine verdeckte Wirklichkeit der Literatur*. München: Hanser.

Heidegger, M. [1929/30] 1983. *Grundbegriffe der Metaphysik*. Frankfurt/Main: Klostermann.

Homann, M. 2001. *Von der Heckerschen Realschule zur Kepler-Oberschule: Berliner und Neuköllner Schulgeschichte von 1747–1992*. Frankfurt/Main: Peter Lang.

Jaeggi, R. 2005. *Entfremdung: Zur Aktualität eines sozialphilosophischen Problems*. Frankfurt/Main: Campus.

Kessel, M. 2001. *Langeweile. Zum Umgang mit Zeit und Gefühlen in Deutschland vom späten 18. bis zum frühen 20. Jahrhundert*. Göttingen: Wallstein.

Kracauer, S. 1977. *Das Ornament der Masse*. Frankfurt/Main: Suhrkamp.

Larkin, B. 2013. The Politics and Poetics of Infrastructure. *Annual Review of Anthropology* 42, 327–343.

Lefèbvre, H. 1974. *Kritik des Alltagslebens*, Vol. I. München: Hanser.

Lefèbvre, H. 1992. *Éléments de rythmanalyse*. Paris: Syllepse.

Lefèbvre, H. 2004. *The Production of Space*. Oxford: Blackwell.

Lehnert, G. (ed.) 2011. *Raum und Gefühl: Der Spatial Turn und die neue Emotionsforschung*. Bielefeld: Transcript.
Löw, M. 2001. *Raumsoziologie*. Frankfurt/Main: Suhrkamp.
Luhmann, N. 1975. Weltzeit und Systemgeschichte. In *Soziologische Aufklärung 2: Aufsätze zur Theorie der Gesellschaft*, edited by N. Luhmann. Opladen: Westdeutscher Verlag, 103–133.
Luhmann, N. 1990. Gleichzeitigkeit und Synchronisation. In *Soziologische Aufklärung 5: Konstruktivistische Perspektiven*, edited by N. Luhmann. Opladen: Westdeutscher Verlag, 95–130.
Macho, T. and Kassung, C. (eds.) 2013. *Kulturtechniken der Synchronisation*. München: Fink.
Martin, D. 2014. Towards a Political Understanding of New Ruins. *International Journal of Urban and Regional Research* 38(3), 1037–1046.
Marx, K. 1973. Entfremdete Arbeit. *MEW*, Vol. 40. Berlin: Dietz, 510–520.
Massumi, B. 2002. *Parables for the Vitual: Movement, Affect, Sensation*. Durham, NC: Duke University Press.
Meyer Spacks, P. 1995. *Bordom: The Literary History of a State of Mind*. Chicago: University of Chicago Press.
Müller, T. 1998. Die Entwicklung des Schulmöbels als Industrieprodukt. In *Das Klassenzimmer. Schulmöbel im 20. Jahrhundert*, edited by T. Müller and R. Schneider. München: Prestel, 9–28.
Navaro-Yashin, Y. 2009. Affective Spaces, Melancholic Objects: Ruination and the Production of Anthropological Knowledge. *Journal of the Royal Anthropological Institute* 15(1), 1–18.
Pease, A. 2012. *Modernism, Feminism and the Culture of Boredom*. Cambridge: Cambridge University Press.
Reckwitz, A. 2012. Affective Spaces: A Praxeological Outlook. *Rethinking History: The Journal of Theory and Practice* 16(2), 241–258.
Reents, F. and Meyer-Sickendie, B. (eds.) 2013. *Stimmung und Methode*. Tübingen: Mohr Siebeck.
Rosa, H. 2009. Kritik der Zeitverhältnisse. Beschleunigung und Entfremdung als Schlüsselbegriffe der Sozialkritik. In *Was ist Kritik?* edited by R. Jaeggi and T. Weschke. Frankfurt/Main: Suhrkamp, 23–54.
Rosa, H. 2012. Geworfen oder getragen? In *Weltbeziehungen im Zeitalter der Beschleunigung*, edited by H. Rosa. Berlin: Suhrkamp, 374–413.
Rosa, H. 2013. *Beschleunigung und Entfremdung*. Berlin: Suhrkamp.
Schell, D. 1999/2000. Zeit in volkskundlicher Perspektive. *Rheinisches Jahrbuch für Volkskunde* 33, 7–14.
Schmidt, M. 2002. Warum Mittelmaß? Deutschlands Bildungsausgaben im Vergleich. *Politische Vierteljahresschrift* 43(1), 3–19.
Schmitz, H. 2007. *Der Leib, der Raum und die Gefühle*. Bielefeld/Locarno: Sirius.
Schneider, R. 1998. Die Suche nach dem idealen Schulbau im 20. Jahrhundert. In *Das Klassenzimmer. Schulmöbel im 20. Jahrhundert*, edited by R. Schneider and T. Müller. München: Prestel, 47–61.
Star, L. 1999. The Ethnography of Infrastructure. *American Behavioral Scientist* 43(3), 377–391.
Stoler, A. 2008. Imperial Debris: Reflections on Ruins and Ruination. *Cultural Anthropology* 23(2), 191–219.

Stoler, A. 2013. 'The Rot Remains': From Ruins to Ruination. In *Imperial Debris: On Ruins and Ruination*, edited by A. Stoler. Durham, NC: Duke, 1–35.

Street, A. 2012. Affective Infrastructure: Hospital Landscapes of Hope and Failure. *Space and Culture* 15(1), 44–56.

Svendson, K. 2002. *Kleine Philosophie der Langeweile*. Frankfurt/Main: Insel.

Toohey, P. 2011. *Boredom: A Lively History*. New Haven, CT: Yale University Press.

Urry, J. 2000. Time. In *Sociology beyond Societies: Mobilities for the Twenty-First Century*, edited by J. Urry. London and New York: Routledge, 105–130.

Wellbery, D. 2003. Stimmung. In *Ästhetische Grundbegriffe*, Vol. 5, edited by K. Barck. Stuttgart: Metzler, 703–733.

Wellgraf, S. 2012. *Hauptschüler: Zur gesellschaftlichen Produktion von Verachtung*. Bielefeld: Transcript.

3 The making of pub atmospheres and George Orwell's *Moon Under Water*

Robert Shaw

Introduction

It is a Saturday in February – the day after I've signed the contract committing me to writing this chapter, in fact. A sunny winter's day, I've headed out for a coastal walk and have arrived in the centre of Sunderland, a city in the North-East of England a few miles from my home. My plan is to get some food before going to the football match that I've got tickets for. I've got slightly more time before the game than can be filled by just eating lunch, so I decide to try and find somewhere which is showing the televised lunchtime football match to keep me occupied for a couple of hours. I'm familiar with the city, but not a regular here, so I don't have a particular venue that I know to visit – instead, I head towards a part of town where I know there's a few pubs, and I walk into the first that might be suitable. A wave of warm air hits me and my glasses quickly steam up. Wiping them on my jumper I look around. The pub's got what I've been searching for – the football is on the telly and menus on a few tables confirm to me that they're serving food. I look around – there are a couple of free tables, but it's busy, and several people are standing and watching the match. I'm a bit uncertain – I know that town will be busy, but I'm hoping for something a bit quieter. I sniff – a stale, beery, sweaty musk enters my nose. It's the straw that's broken the proverbial camel's back: I don't want to eat in an environment like this. I head back out into the street to try another pub.

On the day of the event described above, I ended up spending about 15 minutes wandering between pubs and bars, before finding somewhere that I was happy to eat. Some places were too busy; others did not have the football on TV or were not serving food; and a couple of others were just a bit unpleasant. I realised, as I sat down having finally ordered my lunch, that what I had been looking for was exactly the topic of the chapter that I had just agreed to write: a pub with the right atmosphere. This should not be a surprise. As George Orwell succinctly summarised, "If you are asked why you favour a particular public-house, it would seem natural to put the beer first, but the thing that most appeals to me about the *Moon Under Water* is what people call its atmosphere" (Orwell 1946). I had spent my time in Sunderland judging and evaluating pubs based upon their atmosphere. Orwell's essay from 1946, describing his ideal English pub, focuses on the multiple different features that create the pub's atmosphere, which for him was the key characteristic in judging a public house. As Orwell describes in his essay, I had been looking at a multitude of different features which constituted these atmospheres. I had been sensing the bar spaces themselves, the people and objects within them, and the overall comfort that I felt based upon these features. A subjective, difficult-to-grasp set of criteria had been at play in my mind, and these were not the same criteria that I would have applied on different days, at different times, or at different moments in my life-course. Had I gone into the pubs with different aims, with less time available, or had a different person gone into the pubs, the atmospheres which seemed threatening, rough or boisterous, would have appeared as inclusive, playful or lively. Or simply not have imposed themselves on my senses at all.

Researchers attempting to understand atmosphere have to deal with these difficulties. The object of our study is a personal one, yet one which is created intersubjectively; it is one which requires a reflexive exploration of our positionality, yet one which cannot be explained by a move towards a focus on 'big' social identities. Rather, our experience of atmosphere lies between our individuality and our sociality (Bille et al. 2015). Each individual experiences their own interpretation of the atmosphere of a given space, interpretations which may be contrasting and contradictory but which are nonetheless equally real. Furthermore, elements or even the totality of an atmosphere may be missed, or misunderstood. Despite this, however, atmosphere is central to the alcohol and leisure industry. If Orwell was the first to find the language to describe the importance of atmospheres to pubs and bars, he was neither the first nor the last to notice its centrality. In the UK, the pub industry has been transformed since the mid-1980s by the spread of "pub-cos" (pub companies), chains of pubs and restaurants which have come to own the majority of businesses. The largest of these is "JD Wetherspoons", with over 900 branches at the time of writing, whose pubs were originally designed to replicate the pub as described by George Orwell (Moody and Turner 2013). Atmospheres have received attention then because they are on the one hand individual, unknowable and ephemeral, but on the other predictable and universal enough to form the core part of the strategy of highly successful businesses.

This chapter follows Thibaud's suggestion (discussing the related Francophone concept of ambiance) that the question to ask about these concepts is not purely ontological (what is an atmosphere?); rather, it must also engage with the epistemological: "what does an ambiance make it possible to be, to experience, to do, to perceive and to share?" (Thibaud 2015: 40). Ethnographic methods, I argue, are ideal to answer this question because of their orientation towards experience and practice. Here, the subjectivity of atmospheres stops becoming a problem and starts becoming an opportunity to "grapple with the challenge of sharing empirical narratives that make sense . . . while simultaneously underscoring the situatedness, partiality, contingency, and creativity of that sense-making" (Vannini 2015: 318). Building on Vannini's description of non-representational ethnographies, this chapter places Orwell's fictional account in dialogue with a series of moments from ethnographic work in bars and clubs around the North-East of England, which has formed part of a wider project exploring the atmospheres which constitute the night-time economy (Shaw 2014, 2015). In doing so, it looks to Orwell as a theoriser and describer of bar atmospheres, which have – in comparison to regulation (Talbot 2007), social relations (Jayne, Valentine et al. 2011) and more recently the diversity of actants involved in consumption of alcohol (Demant 2009; Bøhling 2015) – received relatively little attention in contemporary alcohol or night-time economy studies. The core argument is that bar spaces are best understood as spaces in which both material and immaterial elements are constantly being enrolled to produce consumption-oriented atmospheres that produce the wide variety of bar spaces.

Understanding bar spaces

Working class drinking cultures in Orwell's Britain

Globally, drinking cultures are highly varied. In a European context, Järvinen and Room (2007) argue that despite changes in a more fluid, heterogeneous global era, drinking cultures can still be distinguished by broad regional trends based upon predominant alcoholic beverages consumed (Northern beer drinking, Mediterranean wine drinking, and Eastern/Nordic spirit drinking). On a smaller scale, a 2012 Joseph Rowntree Foundation report concluded that local drinking cultures within the UK showed significant variation (Roberts, Townshend et al. 2012), findings which fit in with quantitative research that has revealed a variation in the typology, amount and frequency of drinking habits and attitudes towards drinking between the UK's regions (Shelton and Savell 2011). While the details in this review and chapter focus on the British experience, the underlying arguments – that the control of spaces of (public) alcohol consumption might be best understood through the concept of atmosphere – is more broadly applicable. Within in the UK, licensed public drinking establishments form one of the key sites at which practices of alcohol consumption, production and regulation intersect with broader questions over the governance and contestation of urban public space.[1] Indeed, the traditional name for drinking establishments, the 'public house' reveals the core tension that is at the heart of bar drinking in the UK. On the one hand, the 'public' in this name reveals a building that is notionally open to all – a 'public space'. On the other, it is a private space, a house: closed off from the city at large, with traditional frosted glass preventing the flow of information between inside and out. Pubs have thus long been associated with contestation over access, and forms of exclusion and inclusion enacted both through legislation but also through social practices. Such exclusions have been associated with big social groupings such as gender, age, class and race but also with other more nuanced forms of social difference. Crucially, atmosphere helps us access in particular the diverse actants involved in producing this second form of social exclusion, without losing sight of key questions of governance. At its heart, traditional bar drinking in the UK has created spaces and atmospheres which reflect this contested culture of public/private.

The Mass-Observation project summarised the centrality of bars for working class culture in 1940s Britain:

> Of the social institutions that mould men's [sic] lives between home and work in an industrial town, such as Worktown, the pub has more buildings, holds more people, takes more of their time and money, than church, cinema, dance-hall and political organisations put together.
>
> (Mass-Observation Project 1970: 17)

The detailed ethnography of Mass Observation revealed the pub to be not just a place of alcohol consumption, but as a key site for a series of social practices. Savings clubs, trade unions and other political organisations, a variety of legal and illegal economic activities and broader socialisation are revealed to take place in the pub. Sitting in the public/private tension described above, the pub might be labelled as 'third-place' (Oldenburg 1999), though my preference is to avoid the addition of this category and instead to reveal and explore the tension signified by the '/'. As Vasey argues, the British pub has traditionally been a "sociopetal" space, which favours interaction across the whole room of people attending, with a tradition in particular of pub 'regulars' whose interactions might consist only of meetings in the bar itself (Vasey 1990).

Returning to Mass Observation, the drinkers identified in pubs reflect the complexity of users of any hybrid public/private space. On the one hand, "there are pub-goers amongst both sexes [sic], amongst all adult age groups . . . we do not find them to be restricted to any special type of occupation . . . nor are they restricted to any one social class" (Mass-Observation Project 1970: 154). This reflects other work which has revealed that, building on their 'public' nature, there has always been variation within the populations who use pubs (Beckingham 2012). However, and drawing from the role of the private in the pub, the same studies have also noted that within this heterogeneity, bars have had significantly unequal levels of use and access. To summarise, the pub has historically been a space of white, working-class male culture, with middle-class drinking present but more commonly associated with the home (a trend noted by research in both the 1940s and the 2000s) (Mass-Observation Project 1970; Holloway, Jayne et al. 2008). At the time of the Mass Observation project, women were to be found in pubs, although they were usually excluded from certain areas or might attend less frequently (Mass-Observation Project 1970). Indeed, the inclusion of women can be understood as an inclusion entirely on the basis of their inclusion as second rate, less valid actors within these spaces; in this way, Hey argues that the pub acted as a key replicator of patriarchal culture among working-class communities (Hey 1986). Issues of race are left undiscussed in the Mass Observation research, although some differences on religious grounds are been noted. In exploring this series of inclusions and exclusions, then, a turn to atmospheres of bars presses us to explore how the atmospheres may seem unwelcoming or inviting.

Changing the atmosphere: diversification and feminisation of drinking spaces

Since this ethnographic research in the 1940s, drinking establishments have undergone a series of transformations in the UK. These changes have shaped ever-evolving dominant pub atmospheres. An overview of this period shows a gradual easing of licensing restrictions, and an associated diversification in the style of drinking establishments, the type of people drinking in them, the types of alcohol consumed – all of which are productive of atmosphere (Vasey 1990; Hadfield 2006; Nicholls 2009). A major change for the atmosphere of bars came in 1989, when the Department of Trade and Industry mandated that large breweries – which had previously owned 75 percent of all pubs in the UK – had to sell a portion of their stock. A large proportion of sold pubs were purchased by investment and retail companies which later became 'pubcos'. Whereas pub-owning breweries had specialised in selling their own beers, pubcos were interested in selling a range of food, alcoholic and non-alcoholic drinks. Pubcos diversified: creating distinct brands (*Slug and Lettuce*, *Walkabout*, *Pitcher and Piano, Yates's*); focusing on food and soft drink sales during the day; blurring the lines between pub and club; and creating what has been labelled as a 'night-time high street' (Hadfield 2006). While these changes were hoped to contribute towards a new 'night-time economy' that would revitalise city centres, in practice the night-time high street that emerged consists of nationwide chains of bars selling higher volumes of alcohol to the "mass volume vertical drinker" (Hadfield, Lister et al. 2001). This has been understood as a neoliberalisation of the night-time city and the alcohol market (Shaw 2010). While it is highly contested as to where these changes fall between polls of, on the one hand, producing a series long-term threats to the inclusivity of urban life and the general health of the population, and on the other hand, of a continuation of long-held leisure practices which are demonised by a media focus on particular subgroups (Jayne, Gibson et al. 2012), they have most definitely transformed the variety of atmospheres encountered in different bar spaces.

However, there has been comparatively little focus on what such changes have meant to bar spaces themselves (Hadfield 2006), and it is these changes which are productive of atmosphere. What research does exist suggests that bars and pubs have become more open to a wider audience, as the types of drinking establishments have diversified. In particular, alcohol consumption in pubs has been 'feminised', both with bars made more open and appealing to women and through shifts in the types of drinks consumed. Nonetheless, women in drinking spaces remain subject to a sexualised gaze: "women going out must negotiate the highly gendered and sexualised spaces of the patriarchal night-time economy and their bodies within them" (Waitt, Jessop et al. 2011: 261). On the one hand, women in pubs and bars are expected to have fun, to be available for conversation, flirtation, dancing and sex; on the other, women remain subject to negative discourses surrounding this behaviour in pubs in ways that men are not (Eldridge and Roberts 2008). Similarly, while diversification in drinking establishments have resulted in a generally more open atmosphere, bars still remain sites of racial exclusion, with the imagined night-time economy a predominantly white space (Talbot 2007). While licensing restrictions have become superficially more relaxed, a variety of different controls remain. As Beckingham summarises in his discussion of female drinking in nineteenth century Liverpool, the "licensing process helped shape the social norms around alcohol in public and private space, in and beyond licensed premises" (Beckingham 2012: 662). Similarly, contemporary bar spaces show this governance-practice dialectic relationship. It could be argued that the transformation of pubs in the UK since the 1990s has thus mirrored the transformation of public space: while spaces have become superficially more open and accessible, there remain a series of socially-coded rules, governmental surveillance, and forms of exclusion which work to the detriment of the least powerful (Minton 2009).

British bar spaces have thus been understood as key sites which have acted as a public/private hybrids, and which have produced similarly contrary atmospheres. The changes that have occurred to them since the Mass Observation project offered a first detailed ethnography of pub culture have been generated by a social liberalisation and an economic neoliberalisation. On the one hand, the masculinist, white, narrow culture associated with the pub has been broadened; on the other, forms of control and governance have persisted which now control behaviour less on the basis of identity and more on the basis of subjectivity and behaviour. Throughout, this question of 'atmosphere' has been haunting these understandings of the bar. As pubs have been modernised, atmospheres have become more inclusive and open, but also more oriented around the consumption-purpose of the pub as opposed to the social role identified by Mass Observation.

Elsewhere, I have described atmospheres as "a geographical phenomenon in which a particular assemblage 'gains place'" (Shaw 2014: 88). Atmospheres are emergent from place because, as this chapter outlines, they are resultant from the specific arrangements of objects, bodies, affects and images present in a location. Here, affect takes that double meaning that has been developed by Deleuze and scholars who have since referenced him, as first the embodied experience of the ways in which other 'bodies', broadly defined, shape or modify (that is, 'affect') us (Deleuze 1978), and also as a name for the collective forces which these interactions cause (such as 'hope', 'fear', etc., see Berlant 2011). My perspective follows strongly from de Certeau, who has described how certain places become dominated by practices which have their 'proper place' (de Certeau 1984), coming to control and characterise those spaces. This hints at the idea certain places might be more atmospheric than others (Edensor 2012): building on both de Certeau but also the work of Casey, Duff describes these places as having 'thicker' atmospheres: such places "support more intensive, or affectively resonant, experiences of [affective] outgoing and incoming" (Duff 2010: 886). We might understand this as a description of certain places having more intense and/or more strongly cultivated atmospheres. The mixture of deeply ingrained cultural practices and strongly contested forms of culture as revealed by academic understandings of bar spaces contribute to their becoming places of particularly thick atmospheres. The changes outlined above suggest that these thick atmospheres have, since the

1940s, been mixed and stirred in a variety of ways. In the next stage of this chapter, I draw from my own ethnographic data and Orwell's *Moon under Water* text. In so doing I drill down into the variety of different actants that produce atmosphere, before focusing on the orientation towards consumption.

Understanding the multiple moons under multiple waters

Space and materiality

> To begin with, its whole architecture and fittings are uncompromisingly Victorian. It has no glass-topped tables or other modern miseries, and, on the other hand, no sham roof-beams, ingle-nooks or plastic panels masquerading as oak . . . everything has the solid, comfortable ugliness of the nineteenth century.
>
> (Orwell 1946)

The pubs and bars of Newcastle have a varied range of bar architecture. The city's swankiest clubs and bars can be found along Collingwood Street – an area called 'the Diamond Strip'. Here, bars and clubs are situated in buildings not originally designed as leisure venues, but converted during the 1990s from previous uses. Many are former shops or professional offices, but the largest, which I visited several times during my research, is in an old bank ("My bank is now a trendy wine bar" as the 1990s UK advert had it). The bar runs long, along one side of the large, open room that forms the bulk of this bar. Booths are located on the side of the room opposite the bar, and above these is a mezzanine level with room for standing. At the end of the room is a second raised level with tables and seating. The bar is dimly lit, with the furnishings and carpets largely red. Lights rotate around the bar, sending different patterns in time to the music. This bar, like many others, is full of screens – during the day these play music videos, but at night they show images of clubbers and promote drinks offers. While the building architecture from the 18th century is older even than Orwell's "uncompromisingly Victorian" Moon under Water, the fittings are much more 21st century.

Other modern bars have subtle variations on this theme. One of the bars in the city's more mainstream entertainment area, the Bigg Market, is fitted out in almost entirely silver décor – this 90s-themed bar is going for cheese and kitsch – the music is ridiculous, and the club is too. Others are built to enhance the visibility of dancers, either from members of the public or staff employed by the bar. One bar in 'The Gate' entertainment complex has what appears to be a full replica of Bentham's panopticon, although here brightly lit dancing staff in the middle are always visible – it is the watching crowd, hidden around them, who may or may not be observing. Another places a stage in a second-floor window – female customers are encouraged to dance here, the bar relying on the customers themselves to attract others in. And Newcastle does have its fair share of Moon under Waters – popular 'old man' pubs, where the décor would have looked old fashioned in Orwell's time.

The materiality of a bar's architecture and internal design is perhaps the starting point for the atmospheres which emerge. Here, my use of materiality derives from a tradition which has sought to integrate the role of materials into practice-based theories (Ingold 2005; Anderson and Wiley 2009; Miller 2008). Such approaches do not seek to invoke materiality as 'physicality'. Rather, they view all interactions as imbued with important material qualities. As such, "matter potentially takes place with the capacities and properties of any element (i.e. earth, wind, fire, air) and/or any state" (Anderson and Wiley 2009: 319). Such an understanding of the productive role of materials in atmosphere can be gleamed from Orwell. Building design might not dictate the activities that take place in a location, but it does delimit the field of possible behaviour (Hadfield 2006). Small, cramped pubs will not be able to create the space for large dancefloors – pubs which try this typically quickly revert to type. More modern constructions are – particularly in the city's chain pubs – meticulously designed to produce a series of affective encounters and a particular atmosphere. Cache (1995) refers to this as "the frame of probability" of architecture (ibid.: 24), which he divides into two key features. The first is 'separation', which in practice both divides spaces but also allows for people to be brought together as well. The pubs that Orwell and the Mass Observation project describe both consist of multiple divided rooms; this contrasts with the open spaces of the bar described above (Orwell 1946; Mass-Observation Project 1970). The opening up of bar spaces, and the reduction of separation, is one of the ways that modern bars have attempted to create more public atmospheres: although not universally true (recall that we are talking about 'frames of probability), spaces which are more open tend to have thinner atmospheres (Edensor 2015), as affects are more easily able to diffuse or escape. If thick bar atmospheres emerge from more densely separated sites, then the second role of architecture – the 'selection' – starts to shape the type of atmosphere that we find. As Cache argues, architecture first "removed us from the territory; the second function re-establishes connections, selectively" (ibid.: 24). Here, Cache argues that this function of architecture is represented by the window, which selectively allows light through and helps cast a series of different shadows internally. Affective selection still goes on in bars and pubs of course, and this greatly shapes the atmosphere and as such whether a bar is welcoming or off-putting. This leads us onto the many immaterial sources of atmospheres in bar spaces.

Space and immateriality

> In the Moon under Water it is always quiet enough to talk. The house possesses neither a radio nor a piano, and even on Christmas Eve and such occasions the singing that happens is of a decorous kind.
>
> (Orwell 1946)

My research in Newcastle consisted of a series of short participant observation periods with different actors in the night-time economy, interspersed with periods of 'deep hanging out'. On a night's walk around the city centre, I would (usually unaccompanied) spend a period of two to four hours following a habitual series of routes around the city. Orwell identifies noise as a key element of bar atmospheres. Unlike Orwell's *Moon under Water*, most pubs and bars in the city centre do play recorded music. The large bar that I described previously played mainly contemporary chart music, a common feature of most pubs and bars in the centre of Newcastle. The music set at a level which is just about tolerable for two people to chat: I get the sense that they want to allow for people to sit and talk – the atmosphere has to be inviting – but to encourage drinking and dancing. At this bar, a few groups dance though there is a slightly empty space in the middle of the bar: people seem a little uncertain about creating a 'dancefloor' in this space. Other pubs are more interested in facilitating 'lingering', and here music levels might be lower. Lighting varies significantly between bars, though there appears to be no simple correlation between particular lighting levels and particular atmospheres. In other words, the effect of lighting is dependent upon the relationship between the light and the materials on which it falls (Bille 2015).

Ingold argues that what we call the immaterial is in fact "the medium through which persons and organisms move in perception and action" (Ingold 2005: 97). He rejects the distinction between materiality and immateriality, or at least rejects that these are *oppositions*. While the analytical distinction between the material and immaterial can be useful, perspectives on atmosphere have also rejected this distinction (see for example McCormack 2010). Indeed what emerged from my ethnographic work, and through a reading of *Moon under Water*, is the very material sources of the ways in which immateriality shaped the bar space. Lighting and sound levels were particularly key. Interestingly, both sound and light operated similarly, reverberating around bar spaces and changing atmospheres but with no inherent relationship between greater or lesser levels. In some pubs, quiet produces an intimidating atmosphere, particularly for people who do not fit into the spaces' dominant social groups; in other locations, a quiet pub can create a welcoming atmosphere that attracts groups searching an escape from the noise of other venues. Lighting worked similarly, with no inherent relationship between atmosphere and darkness/brightness of a bar. Here, then, it is suggested that the effect of the immaterial depends upon its 'fit' with the material; the atmosphere is emergent from the relation between the two.

One notable absence from the bar spaces of my ethnography was another 'immaterial' element, that of smoke. Any reflection on bar atmospheres in the UK prior to the smoking ban might have noted the major effect that the presence of a layer of cigarette smoke has on the atmosphere of bar spaces (Tan 2012). In many countries, the smoky atmosphere of bars is a key element of drinking spaces. Debates over the smoking ban revealed the importance of the immaterial in making bar spaces accessible or exclusive, and the complexities embedded within this. On the one hand, the presence of smoking has been associated with a particularly male environment, discouraging the opening up and cleaning of pubs which has helped make spaces more accessible. On the other, these atmospheres have been praised as inclusive; they are certainly, literally, 'thick', encouraging and affinity with the pub. Accounts of atmosphere need to deal with the reality that the smoky bar may be simultaneously attractive and repulsive, and the smoking ban's sanitisation both includes and excludes.

In the two previous sections, we have considered the emergence of atmosphere from both material and immaterial sources, and focused on the 'fit' between these. I want to move on now to consider the power and control over this emergence, focusing on consumption practices.

Atmospheres and consumption

"The barmaids know most of their customers by name, and take a personal interest in everyone. They are all middle-aged women – two of them have their hair dyed in quite surprising shades – and they call everyone 'dear', irrespective of age or sex. ('Dear,' not 'ducky': pubs where the barmaid calls you 'ducky' always have a disagreeable raffish atmosphere)" (Orwell 1946).

As part of my research I attended a hiring day at one of the city's clubs. Following an informal presentation in which we were introduced to this particular chain's "unique customer focused values", we were set a series of individual and group tasks during which staff monitored our performance. In the opening activity we passed a beach ball around the group – on receiving it we had to give our name, our favourite movie, and an interesting fact about ourselves. My favourite film ('24 Hour Party People') received largely blank stares, and my interesting fact ("I once carried a live goose on a bus around Newcastle") probably required more explanation to make any sense. More popular answers seemed to focus on children's and cheesy films from the 1980s and 1990s ('Back to the Future' and 'Lion King' received the biggest group acclimation), and tales from package holidays. Our next activity was a group exercise – in teams of five our challenge was to create and perform a short sketch. Our performances needed to include the words 'giraffe', 'peanut', 'helicopter' and the club's name – humour based on attempts to indicate the fun-loving and 'random' side of working at the club. Groups were encouraged to put on performances that were loud, unusual and involved lots of action. These activities, by the way, were for a series of part time and temporary bar serving or glass collecting jobs, paying only slightly higher than the UK's minimum wage.

These tasks were testing the communicative abilities of the prospective employees – to make a group of strangers laugh and understand us in a quick piece of conversation. As such, this was part of attempts to foster a workforce whose corporeal capacity for interaction and being with others is their main tool of production (Virno 2007), that is, those people who had the greatest capacity to affect others (Deleuze 1978). However, staff were being tested not only for their affective capacities, but also for their ability to read and judge an atmosphere in order to encourage greater consumption – to add to or replenish the thickness of the atmosphere (Duff 2010). Orwell's comments show a long history to this relationship between corporeal capacity and the affective atmosphere of a bar. Indeed, the Mass Observation project similarly quotes a publican who states that "the good landlord is of a type who would lead a pack; . . . leads the easily led. He along with his wife . . . are a type. Grotesque in the lady, coming out with flashy dress and speech in the more choice specimens" (Mass Observation Project 1970: 52).

While emergent from the material and immaterial features of a space, these atmospheres do not emerge by chance. Rather, bar atmospheres are constantly being shaped by actors who may not have the capacity to control the atmosphere but who nonetheless complete several strategies to try and shape the emergent atmospheres.

While this use of corporeal capacity suggests a quick and responsive way of producing atmospheres, timetables and product choices also excrete atmosphere over longer timeframes:

> You cannot get dinner at the Moon under Water, but there is always the snack counter where you can get liver-sausage sandwiches, mussels (a speciality of the house), cheese, pickles and those large biscuits with caraway seeds in them which only seem to exist in public-houses. Upstairs, six days a week, you can get a good, solid lunch – for example, a cut off the joint, two vegetables and boiled jam roll – for about three shillings. The special pleasure of this lunch is that you can have draught stout with it. I doubt whether as many as 10 per cent of London pubs serve draught stout, but the Moon Under Water is one of them. It is a soft, creamy sort of stout, and it goes better in a pewter pot.
>
> (Orwell 1946)

Across all of Orwell's description, and pervading throughout my ethnographic work, is the importance of the commercial side of bar spaces, as exemplified by Orwell in the food and beer available. As noted, Orwell's description was heavily influential on the development of the *JD Wetherspoon* chain of pubs, which by 2010 had a turnover of £1billion and which has over 900 branches in the UK. Inspired by Orwell, the chain seeks to "control the atmosphere" (Moody and Turner 2013: 257) of its pubs in order to produce a standardised, familiar product which maximises consumption by changing and moulding atmospheres over the course of the day. In Newcastle city centre there are five *Wetherspoon* pubs, two under its *Lloyds* sub-brand. It's in one of these where I explored the regimented timetabling of night and day, as atmospheres are shifted and shaped to encourage different forms of consumption. On a March Saturday in mid-afternoon, the bar is full of people taking a break from weekend-activities: families or couples with shopping, people meeting for tea and coffee. Pop music plays in the background, and a wide range of products are being consumed. At 16:20, the pop music that had been playing throughout the day was made notably louder. Staff began moving around the bar, and five minutes later we were asked to move from our stools, in the centre of the bar, to seats elsewhere in the bar; this allowed for a large central dancefloor to be opened up. No new family groups arrived after this point, and the profile of drinks being consumed shifted towards alcohol. The music was made louder again at 16:45, and became disassociated from the video screens, which had previously been playing music videos to accompany the sound. At 17:00 the food menu changed with fewer options being offered, and the dancefloor was fully cleared. Shortly after this the lighting of the bar was changed, with red lighting replacing the daytime white light at around 17:30. The final major change in the bar itself came at 19:00, when a DJ took over from the loop of pre-recorded music. During this period, the profile of consumers in the bars gradually became younger, and groups began to dance, shout and drink as on a night out.

As well as hiring staff to shape the atmospheres through their affective capacities, both the Moon under Water and *Wetherspoon's* are also generating atmosphere through more regimented regimes of spatial practice. Again, this reveals the ways in which bar atmospheres – and many atmospheres more broadly – are always controlled, emergent from a series of strategies and tactics, to use the vocabulary of de Certeau, which attempt to assert their own agenda upon a place through the creation of a dominant atmosphere.

Conclusion: Orwell's embedded atmospheres

Among the various atmosphere theorists that have been drawn upon in the literature – Böhme, Deleuze, Sloterdijk, de Certeau and Brennan amongst others – Orwell might seem an unusual addition. This chapter has been an attempt to use his work to offer a typology of how atmosphere is produced from a mixture of material and immaterial features. His essay *Moon under Water* offers both an evocative (fictional) ethnographic description of a bar atmosphere, but also a more general typology for how atmospheres might be understood. In particular, 'atmosphere' in Orwell is deeply embedded in place, emerging from a location's material, immaterial and interpersonal relations. Atmosphere is for Orwell excreted out of dozens if not hundreds or thousands of repeated practices. In thinking through atmospheres ethnographically, I am minded of Guattari's claim that:

> Over thousands of years, perhaps in imitation of crustaceans or termites, human beings have acquired the habit of encasing themselves in all kinds of shells: buildings, clothes, cars, images and messages, that they never stop secreting like a skin, adhering to the flesh of their existence just as much as do the bones of their skeletons.
>
> (Guattari 2005: 119)

Ethnographic exploration of atmosphere, seeking to create productive descriptions and narratives of the spaces that atmospheres make possible (Thibaud 2015; Vannini 2015), might best be understood as attempts to capture both the secretion of atmosphere and the experience of being within it. In so doing we see in atmosphere a concept which defies the distinctions between agency/structure, permanence/transience, individual/collective that pervades social science. Rather, atmospheres demand a research method which conveys the multiple practices and actors which shape the 'being-there'.

Bar spaces in the UK are distinguished by their atmosphere. Two different venues can serve the same drinks, play the same music and be in the same street, but have distinctly different atmospheres. This comes from the thickness of atmospheres in most licensed premises: a thickness which both emerges slowly over time, through repetition of a series of learned practices, and also which is reinforced by the actions of staff and others who respond quickly to the emergent affects of a bar or club space. Bars are spaces whose atmosphere is both material and immaterial, but perhaps where most crucially these two are intertwined: atmospheres emerge from their interaction. As public/private spaces, atmospheres become important due to their role in producing the 'soft exclusion' associated with feeling (un)comfortable within an atmosphere. It is important to note that, in this context, atmospheres of bar spaces are not produced neutrally or by accident. While many features may be unintentional, there are always in consumption places attempts to govern, produce and shape atmosphere.

Note

1 Although they are far from the only spaces of alcohol consumption more broadly. Indeed for some social groups or subcultures, drinking in the home (Holloway, Jayne et al. 2008) or in open public spaces (Wilkinson 2015) is more common.

Bibliography

Anderson, B. and Wiley, J. 2009. On Geography and Materiality *Environment and Planning A* 41(2), 318–335.

Beckingham, D. 2012. Gender, Space, and Drunkenness: Liverpool's Licensed Premises, 1860–1914. *Annals of the Association of American Geographers* 102(3), 647–666.

Berlant, L. 2011. *Cruel Optimism*. Durham, NC: Duke University Press.

Bille, M. 20015. Lighting Up Cost Atmospheres in Denmark. *Emotion, Space and Society* 15(1), 56–63.

Bille, M., Bjerregaard, P. and Sørensen, T. F. 2015. Staging Atmospheres: Materiality, Culture, and the Texture of the in-Between. *Emotion, Space and Society* 15(1), 31–38.

Bøhling, F. 2015. Alcoholic Assemblages: Exploring Fluid Subjects in the Night-Time Economy. *Geoforum* 58, 132–142.

Cache, B. 1995. *Earth Moves: The Furnishing of Territories*. Cambridge, MA: MIT Press.

de Certeau, M. 1984. *The Practice of Everyday Life*. Berkeley, CA: University of California Press.

Deleuze, G. 1978. *On Spinoza*. Available at http://deleuzelectures.blogspot.co.uk/2007/02/on-spinoza.html [Accessed 20 November 2013]

Demant, J. 2009. When Alcohol Acts: An Actor-Network Approach to Teenagers, Alcohol and Parties. *Body & Society* 15(1), 25–46.

Duff, C. 2010. On the Role of Affect and Practice in the Production of Place. *Environment and Planning D: Society and Space* 28(5), 881–895.

Edensor, T. 2012. Illuminated Atmospheres: Anticipating and Reproducing the Flow of Affective Experience in Blackpool. *Environment and Planning D: Society and Space* 30(6), 1103–1122.

Edensor, T. 2015. Producing Atmospheres at the Match: Fan Cultures, Commercialisation and Mood Management in English Football. *Emotion, Space and Society* 15(1), 82–89.

Eldridge, A. and Roberts, M. 2008. Hen Parties: Bonding or Brawling? *Drugs-Education Prevention and Policy* 15(3), 323–328.

Guattari, F. 2005. Architectural Enunciations. *Interstices* 6, 119–125.

Hadfield, P. 2006. *Bar Wars: Contesting the Night in Contemporary British Cities*. Oxford: Oxford University Press.

Hadfield, P., Lister, S., Hobbs, D. and Winlow, S. 2001. The '24-Hour City' – Condition Critical. *Town and Country Planning* 70(11), 300–302.

Hey, V. 1986. *Patriarchy and Pub Culture*. London: Tavistock.

Holloway, S. L., Jayne, M. and Valentine, G. 2008. 'Sainsbury's Is My Local': English Alcohol Policy, Domestic Drinking Practices and the Meaning of Home. *Transactions of the Institute of British Geographers* 33(4), 532–547.

Ingold, T. 2005. The Eye of the Storm: Visual Perception and the Weather. *Visual Studies* 20(2), 97–104.

Järvinen, M. and Room, R. 2007. Introduction to Youth Drinking Cultures: European Experiences. In *Youth Drinking Cultures: European Experiences*, edited by M. Järvinen and R. Room. Aldershot: Ashgate, 1–16.

Jayne, M., Gibsonm, C., Waitt, G. and Valentine, G. 2012. Drunken Mobilities: Backpackers, Alcohol, 'Doing Place.' *Tourist Studies* 12(3), 211–231.

Jayne, M., Valentine, G. and Holloway, S. L. 2011. *Alcohol, Drinking, Drunkenness: (Dis) Orderly Spaces*. Aldershot: Ashgate.

Mass-Observation Project. 1970. *The Pub and the People: A Worktown Study*. Welwyn Garden City: Seven Dials Press Ltd.

McCormack, D. P. 2010. Remotely Sensing Affective Afterlives: The Spectral Geographies of Material Remains. *Annals of the Association of American Geographers* 100(3), 640–654.

Miller, D. 2008. *The Comfort of Things*. London: Polity.

Minton, A. 2009. *Ground Control: Fear and Happiness in the Twenty-First-Century City*. London: Penguin.

Moody, P. and Turner, R. 2013. *The Search for the Perfect Pub*. London: Orion.

Nicholls, J. 2009. *The Politics of Alcohol*. Manchester: Manchester University Press.

Oldenburg, R. 1999. *The Great Good Places*. New York: Marlowe.

Orwell, G. 1946. *The Moon Under Water*. Available at http://theorwellprize.co.uk/george-orwell/by-orwell/essays-and-other-works/the-moon-under-water/ [Accessed 10 May 2016]

Roberts, M., Townshend, T., Pappalepore, I., Eldridge, A. and Mulyawan, B. 2012. *Local Variations in Youth Drinking Cultures*. York: Josef Rowntree Foundation.

Shaw, R. 2010. Neoliberal Subjectivities and the Development of the Night-Time Economy in British Cities. *Geography Compass* 4(7), 893–903.

Shaw, R. 2014. Beyond Night-Time Economy: Affective Atmospheres of the Urban Night. *Geoforum* 5, 87–95.

Shaw, R. 2015. 'Alive After Five': Constructing the Neoliberal Night in Newcastle-Upon-Tyne. *Urban Studies* 52(3), 456–470.

Shelton, N. and Savell, E. 2011. The Geography of Binge Drinking: The Role of Alcohol-Related Knowledge, Behaviours and Attitudes. Results From the Health Survey for England 2007. *Health & Place* 17(3), 784–792.

Talbot, D. 2007. *Regulating the Night: Race Culture and Exclusion in the Making of the Night-Time Economy*. Aldershot: Ashgate.

Tan, Q. H. 2012. Towards an Affective Smoking Geography. *Geography Compass* 6(9), 533–545.

Thibaud, J.-P. 2015. The Backstage of Urban Ambiances: When Atmospheres Pervade Everyday Experience. *Emotion, Space and Society* 15, 39–46.

Vannini, P. 2015. Non-Representational Ethnography: New Ways of Animating Lifeworlds. *Cultural Geographies* 22(2), 317–327.

Vasey, D. E. 1990. *The Pub and English Social Change*. New York: AMS Press.

Virno, P. 2007. General Intellect. *Historical Materialism* 15(3), 3–8.

Waitt, G., Jessop, L. and Gorman-Murray, A. 2011. 'The Guys in There Just Expect to Be Laid': Embodied and Gendered Socio-Spatial Practices of a 'Night Out' in Wollongong, Australia. *Gender, Place & Culture* 18(2), 255–275.

Wilkinson, S., 2015. Alcohol, young people and urban life. *Geography Compass*, 9(3), 115–126.

4 Vapours in the sphere

Malaria, atmosphere and landscape in wet lands[1] of Agro Pontino, Italy

Paolo Gruppuso

Introduction

> Along the road, aside me, in front of my eyes (and my ears), a spectacle of an ambiguous beauty appears. From the groves of reeds along the canal banks the smoke rises in vacuous columns. I am driving, and little by little as I near the canal, I perceive stronger and stronger the smell and the increasing crackle, unexpectedly impressive, of the reeds that burn. I stop in order to admire the show: it is beautiful, majestic, ancient.
>
> After a few minutes I restart driving towards the *Gricilli* area. I am completely absorbed in the landscape; I think of the old Pontine Marshes, of their atmosphere and of the fascination of travellers and artists. I try to recognise some of the places painted by the artists travelling in the region before the reclamation, . . . everything has changed, the landscape is unrecognisable: the land has been transformed, trees have grown or have been cut down, the number of houses has increased . . . everything is different, but the atmosphere.
>
> (Personal fieldnotes, 10 March 2012)

This excerpt from my fieldnotes concerns the annual activity of burning reeds in the *Gricilli* area, a protected wetland in Agro Pontino, Italy, where I carried out my ethnographic fieldwork.[2] The practice of burn-beating represents a form of care towards the *Gricilli* wet landscape which, even though in decline, has a long history in the area. Looking into the past of Agro Pontino, i.e. the Pontine Marshes, people used this technique of fire within a complex array of agricultural practices aimed at nurturing that particular wet landscape. Apart from producing vacuous columns of smoke (see Figure 4.1), the practice of burn-beating, when it was widespread in the Pontine Marshes, had the result of consuming the upper layer of peat in the soil. This process produced a remarkable lowering of the land, creating ponds where the surface water stagnated. In specific meteorological conditions this water evaporated in the air, producing vapours that permeated the landscape. In the period between the 18th and the 19th century, these vapours were thought to be poisonous miasmas associated with malaria, at that time understood as 'bad air', a mysterious atmospheric phenomenon affecting marshlands and wet landscapes in general. This understanding recalls the etymology of the term atmosphere from the ancient Greek words *àthmos* (ἄθμος – vapour)

Figure 4.1 Burn-beating in the *Gricilli* area

Photograph by the author

and *sphàira* (σφαίρα – sphere), literally meaning *vapour in the sphere*. These vapours had an importance in characterising a peculiar idea of atmosphere in the region, perceived at the same time as dreadful and charming. This chapter explores the history, meaning and imaginary of this particular atmosphere related to the vanished landscape of the Pontine Marshes between the 18th and the 19th century, compared with my own ethnographic experience of the landscape in Agro Pontino.

The association between atmosphere, malaria and landscape, in the context of my research, is interesting for three main reasons. First, it affords the possibility to explore the complexity of the concept of atmosphere through the lens of physicians, meteorologists and aestheticians in the period between the 18th and the 19th century. Second, it allows a discussion of the notion of atmosphere not only in terms of visual, aural, olfactory or affective qualities but as the totality of the relationships in which living beings are immersed, as the basic constituent of their lifeworld (see Böhme 1993). Third, it challenges the idea that atmosphere is only a medium, a "condition of interaction" (Ingold 2015: 70) or "the environment in which sensorial perception takes place" (Tedeschini 2014: 137), presenting the atmosphere both as an object of perception and interaction and as an artefact, the result of the actual engagement of living beings within a particular environment. In exploring the notion of atmosphere and the role it plays in understanding and conceptualising landscape from an anthropological perspective, this chapter is a contribution to recent debates that seek to clarify the heuristic value of the concept for social analysis (Bille et al. 2015: 32).

After an initial exploration of the often ambiguous concept of atmosphere, this chapter will first present the environmental context of Agro Pontino before the 1930s, when the fascist regime reclaimed the Pontine Marshes, perceived by scientists and humanists as permeated by a "pestilential" and "deadly" atmosphere (e.g.: Clarke 1812: 652; Blodget 1857: 464). In describing the Marshes I also introduce and clarify my understanding of landscape as taskscape (Ingold 1993). Subsequently, I discuss literary as well as scientific historical sources, which illustrate the peculiar understanding of atmosphere in the Pontine Marshes, as associated with malaria, at that time perceived as a combination of vapours, miasmas and 'bad air', while at the same time emphasising the aesthetic allure of the very same characteristic atmosphere. Following a discussion highlighting the continuity between the atmosphere I experienced during my fieldwork in Agro Pontino and the atmosphere of the old Pontine Marshes I will conclude with a reflection on the relations between atmosphere, body and landscape.

The ambiguity of atmosphere

The fieldnotes at the outset describe my encounter with a landscape permeated with something I would call atmosphere: a combination of ephemeral, unspeakable and evanescent feelings with a specific physical phenomenon, namely the transformation of organic materials by means of combustion or evaporation. My experience of atmosphere resonates with Anderson's reflections about what he terms "affective atmospheres" (2009): "On the one hand, atmospheres are real phenomena. . . . On the other they are not necessarily sensible phenomena" (ibid. 78). The same tension is also highlighted by geographer Derek McCormack when discussing the attempt to fly to the North Pole in a hydrogen-filled balloon carried out at the end of the 19th century by a Swedish engineer (McCormack 2008). McCormack notices the complex ambiguity of atmosphere, emphasising both its meteorological and affective qualities (McCormack 2008: 413). The indeterminacy of atmosphere, understood as weather, is also discussed by anthropologist Tim Ingold, who explores the Western divide between the perceived real materiality of the landscape compared to the perceived imaginative immateriality of the weather (Ingold 2005: 103, see also Ingold 2006, 2015).

The concept of atmosphere, then, holds a series of tensions between materiality and immateriality, permanence and ephemerality, everyday speech and specialised knowledge, aesthetic appreciation and meteorological measurements, subjectivity and objectivity. As a result, an uncertainty or indeterminacy in relation to the concept of atmosphere as a subject of inquiry emerges. It is the ambiguity, perhaps, that makes the notion of atmosphere interesting for anthropology, and makes it ready to be put to productive conceptual use, as it allows to see the atmosphere simultaneously as an object of perception, as the result of interaction, as well as an artefact – realising that atmosphere is a process resulting from the engagement of living beings within a particular environment.

Cultivated bogs: the Pontine Marshes and their atmosphere

Agro Pontino is a region located 70 km south of Rome in Italy. It has been affected, especially in the last three centuries, by several attempts of land reclamation. The definitive and most important, named *Bonifica Integrale* (complete reclamation), was conducted during the 1930s by the fascist regime, which drained the Pontine Marshes, perceived as a remote and sparsely inhabited wasteland (Gruppuso 2014). In spite of the fascist propaganda, which depicted a long lasting representation of the Marshes as "nothing but a malaria infested marsh" (Harris 1957: 311), archive materials demonstrate that the Pontine Marshes were an inhabited, rich and productive landscape whose wealth was framed within a peculiar economy based on agriculture, forestry, husbandry, hunting and fishing (e.g. Folchi 1996; Gruppuso 2014).

Italian geographer Federica Letizia Cavallo describes this environment as an "amphibious" context "where the prevalence and the changing nature of the liquid element, determined an unstable structure, variable during the long periods of geomorphological processes, during the seasons' cyclical time, and sometimes even during a single day" (Cavallo 2011: 116).[3] In this sense, the productive and working activities in the Pontine Marshes flowed within a series of possibilities, following the seasons as well as the water's flow. This flow resonates with the idea that human beings are ontological amphibians, animals able to move from one element to another (ten Bos 2009: 74). They are hence able to produce an amphibious economy, switching from fishing to hunting and from farming to gathering. This peculiar economy was effectively summarised by the French intellectual Edmond About who, travelling along the Marshes in the nineteenth century, wrote in his diary: "Even the cultivation of the bogs repays the farmer" (1861: 206).

The idea of cultivating bogs highlights the peculiar relations between people and the amphibious landscape of the Marshes. Rather than defining one particular activity, this expression defines a taskscape (Ingold 1993), an array of agricultural practices aimed at keeping the land wet and nurturing the Marshes. Until the beginning of the twentieth century engineers and scientists considered these agricultural practices as significantly related with the hydrogeological condition of the Marshes and with their atmosphere understood in terms of 'bad air'. The Marshes and their atmosphere were considered part and parcel of the same amphibious taskscape; the result of the actual engagement of the local people within the materiality of their environment. They emerged along with the shape of the land, in the activities that human and nonhuman beings performed throughout their lives. This is a key aspect of my argument: it highlights the idea that atmosphere and the landscape are both part of the material world, and that they can be both understood not as independent objects but as a field of relations in-between human and other beings, things and materials (see Ingold 2005).

The notion of taskscape, I suggest, is particularly useful because focusing on tasks, rather than the visible shapes of the land, it allows a better understanding of the amphibious environment of the Pontine Marshes as composed by a combination of different elements, materials and beings, woven together as a meshwork (Ingold 2007, 2011). Moreover, the notion of task, I argue, involves also the idea of care. The inhabitants of the Marshes, accomplishing their tasks, nurtured the Marshes, keeping them alive through "the cultivation of the bogs" (About 1861: 206). This particular expression, as the notion of taskscape, challenges any cultural or natural reductionist interpretation of the landscape: rather than being just a natural phenomenon, the Marshes were a process that involved human and non-human beings immersed in a field of elements, geological forces and materials.

This aspect strongly emerges from the technical reports of the engineers employed in the land reclamation processes in the 18th and the 19th centuries (see Gruppuso 2014). In their understanding the reclamation of the Pontine Marshes was not difficult because of the region's hydrogeology, but because of their inhabitants, considered guilty of cultivating the bog and keeping the land wet. From these reports the Marshes emerge as an artefact, rather than as a natural phenomenon (see Ingold 2013), created through particular agricultural practices, primarily the use of fire. On the one hand, fire was used in order to clean the fields from the residuals of the old crops and to fertilise the soil with the ashes. On the other hand fishermen used fire to produce deep depressions with stagnant water used as fishponds (Nicolai 1800: 286). A suggestive description of fire, seen as an agent able to shape the Pontine wet landscape, can be found in an anonymous report, written at the end of the eighteenth century, which reads:

> The squelchy soil of the Pontine Marshes it is not solid and compact; for two or three spans from its surface it is rather composed by putrescent reeds, logs, leaves and light earth. . . . If fire goes on this soil, it grazes in it, it slithers in it, it founds its home, it drains the ground making it lower. The lowered soil is lower than the water level and it is again waterlogged.
>
> (quoted in Folchi 2002: 207)[4]

The use of fire, within a wider and complex array of amphibious agricultural practices, was considered to be responsible of the wet landscape of the Marshes and of their 'bad air'. In what follows, I present materials that tell more about the understanding of the relations between 'bad air', stagnant water and putrescent soil; the relations between atmosphere, miasmas and the shape of the land.

Soil and air, landscape and atmosphere

The Pontine Marshes became particularly famous between the 18th and the 19th century. Many scientists explored and studied the Marshes from many different perspectives, such as botany, medicine and hydrogeology, in order to drain them. The Marshes also inspired artists, poets, humanists and aestheticians, including Johann Wolfgang Goethe, Alexandre Dumas, and Horace Vernet, who visited the region travelling along the itinerary of the *Grand Tour* (see Mammuccari and Langella 1999). They were all fascinated and at the same time frightened by the Pontine Marshes, and these feelings concerned primarily one aspect, as we will see below, namely their atmosphere, understood as the quintessence of the Marshes: at the same time terrific and lush, malarial and disordered.

These contradictory feelings were the results of two main reasons. The first is that the physiological mechanism of malaria was at that time unknown. The second is that marshlands were perceived until the first half of the last century as ambiguous places not only in Agro Pontino, but all around the world (e.g. Nicholas 1998, 2013; Breda 2001; Traina 1988; Ogden 2011). These reasons are interrelated: the perceived ambiguity of the Pontine Marshes corresponded to the perceived ambiguity of their atmosphere. This sense of ambiguity, related to the presence of malaria in the Marshes, is still very present in Agro Pontino. Amongst the interviews I collected during my fieldwork with the aim of gathering stories about the vanished Pontine Marshes, I found particularly interesting the ideas expressed by E.S., an elderly intellectual from Terracina, the most important administrative centre of the Pontine Marshes, situated on the coast of Agro Pontino. E.S. was a teacher at the local high school and he has always lived in Terracina where he was a well-known and respected researcher in local history as well as an environmentalist engaged in local politics. When I asked him about the relations between the local communities and the Pontine Marshes, he replied:

> The wealth of the marsh represented a pull factor, because as you know, wetlands are the richest environments of the world with regard to the production of biomass. I mean, in a marsh, the production of vegetal as well as animal biomass, is amazing. There was, of course, also the danger to fall sick . . . this looming malaria, with its recrudescence cycles.[5]

This interview emphasises a key aspect to understand the relation between the Marshes and their atmosphere: the perceived ambiguity of the Marshes, derived from its extreme productivity and from the ways in which a multiplicity of forms of life, including dangerous ones like the malarial mosquitoes, thrived in it.

Nevertheless, when thinking about malaria, it is important to remember that the discovery of the cycle of Plasmodium occurred only in the late 19th century and that even later, at the beginning of the last century, was the relation between a particular kind of mosquitoes and the disease discovered (Crotti 2005).[6] Before this scientific breakthrough, mosquitoes were considered only as an extremely annoying inconvenience of living and travelling in the Marshes. Malaria, the Italian contraction for *mal'aria* (*Aria Cattiva* – 'bad air'), on the other hand, was thought to be related with particular environmental qualities, especially the features of soil and its fertility.

The relation between soil and *mal'aria* has been investigated for a long time, puzzling many scientists (see Corbin 1986). Amongst them was the meteorologist Graham Hutchinson, who noticed that "soils are in general to be suspected of possessing deleterious qualities in proportion to their vegetative fertility. . . . " (Hutchinson 1848: 493). These deleterious qualities concerned one aspect, namely the production of miasmas, of poisonous atmosphere. In his treatise on meteorological phenomena Hutchinson included an essay on this very subject; inquiring the cause of 'bad air' he writes:

> Chemical science has hitherto afforded no information regarding the nature of infectious effluvia. Neither the tainted atmosphere of a fever-hospital, nor

> of a city infected with the plague where hundreds are dying daily, nor the noxious air from putrid marshes, nor that which is contained in apartments where persons have died of infectious disease, has ever yet been found by chemical analysis, to contain any substance deleterious to human life, and to presence of which the phenomenon of infection might be referred. Experiments with a view to such discoveries, and particularly to ascertain the nature of marsh miasma, have been often made but uniformly without success.
>
> (Hutchinson 1848: 433)

It is interesting that in this excerpt, as well as the whole treatise, Hutchinson used the terms effluvia, atmosphere, air and miasma interchangeably, pointing to a certain materiality of these scientific objects, which, in his understanding, permeated particular environments and the bodies of their inhabitants. Moreover, the interchangeable use of these terms, which we consider nowadays as different in meaning, suggests the uncertainty about their origin. This aspect emerges clearly in a paragraph where Hutchinson, arguing about the *Treatments of Patients* affected by *mal aria*, writes:

> The most important rule to be observed and which with slight modification, is susceptible of an extended application to all diseases arising from an infected atmosphere, is to remove the patient as soon as possible from the seat of contagion.
>
> (Hutchinson 1848: 487)

In this excerpt, Hutchinson mentions the idea of a "seat of contagion", namely particular environments characterised by an "infected atmosphere". Indeed, at that time, malaria, the 'bad air', was understood as a phenomenon related to meteorological factors like winds and clouds, humidity and dryness as well as with the shapes of the land, mainly with wet landscapes (see Hutchinson 1848: 439).

Indeed, many of the scientists working in the region in the 19th century and before were interested in understanding the relations between the landscape and its atmosphere understood as 'bad air'. In order to illuminate this phenomenon, they looked particularly at the position of woods and water bodies with regard to the villages and to wind directions as well as at the use of water and forests by local communities (Di Tucci 1882). The outcomes of these studies were always pretty contradictory; nevertheless they are interesting because they inscribed the 'bad air', hence the atmosphere, within the frame of relations between local communities and their environment. The 'bad air' of the Marshes was understood as the outcome of the local agricultural practices, namely the "cultivation of the bogs" (About 1861: 206), which was supposed to foster a 'pestilential atmosphere'.

In this interpretation one can sense the long lasting legacy of Hippocrates (see also Corbin 1986: 13). In his treatise *On Airs, Waters and Places*, the Greek physician argued about the relation between marshes and particular kinds of diseases, amongst them *quartan fever*, i.e. malaria, due to "stagnant water from marshes and lakes. . . . " (1978: 152). Putrefaction, decay, and transformation of living materials were considered to affect the atmosphere and to be responsible for 'bad air'. This point is highlighted by Angelo Secchi (1865) who, describing the marshlands around Rome and the Pontine Marshes, wrote that: "The corruption of animal and plant materials in these waters, under the scorching sunbeams in our clear atmosphere, is extremely rapid and it is a fecund source of miasmas" (Secchi 1865: 13).[7]

These miasmas, were understood as "material substances" (Hutchinson 1848: 455). This particular reading of the Marshes' atmosphere, I suggest, questions the opposition between the concrete earthliness of the land and the immateriality of the sky highlighting an idea of the landscape as a "fluid and becoming entity" (Edensor 2010: 230) where the land is not just an interface between earth and sky, but is a "vaguely defined zone of admixture and intermingling" (Ingold quoted in Edensor 2010: 230).

In this sense, the Marshes were considered as a terrestrial concretion of their atmosphere, heterogeneous aggregates of different substances and materials revealed in the aerial medium through particular signs, primarily vapours (see also Corbin 1986: 13).

"The pontine marshes sent up a mysterious steaming vapour"

Michael Taussig illustrates the relation between miasmas, vapours and marshlands, tracing the meaning of the term miasma:

> Miasma: from the Greek, meaning a contagious and dangerous pollution as due to transgression of a supernatural sanction, and more recently meaning an infectious or noxious emanation; "a vapor as from marshes, formerly supposed to poison the air" as *Webster's Dictionary* would have it; emanations from rotting matter of noxious particles causing malaria, as the OED has it.
>
> (Taussig 2004: 175)

The notion of miasma as vapour (see also Corbin 1986), today completely obsolete, resonates with the etymology of the word atmosphere, vapour in the sphere. This notion is interesting because it makes evident the relations between atmosphere, body and landscape: vapours rise from the soil, permeate the atmosphere and seep into human bodies. Many of the aestheticians who travelled along the Marshes between the eighteenth and the 19th centuries focused their descriptions on the vapours that permeated the Marshes, emphasising the materiality of atmosphere, in relation to body and landscape.

This aspect is emphasised by Goethe, who travelling through the Pontine region during February 1787, notes in his diary: "[. . .] we were strongly enough reminded of the danger of the atmosphere, by the blue vapour which, even in this season of the year, hangs above the ground" (Goethe 1817: 171). Goethe points out that the atmosphere, and particularly its danger, is visually discernible as bluish vapour floating above the ground. A similar image emerges from the novel *Le*

Chevalier de Saint-Hermine by Alexadre Dumas (1869), where the plot of one chapter is staged in the Pontine Marshes:

> As they neared the Pontine Marshes, large streaks of vapor, more mists than clouds, rose intermittently toward the sky and passed over the face of the moon like a widow's black gauze veil. And the sky would take on strange unhealthy, yellowish tones.
>
> (Dumas 2007 (1869): 657)

In this passage Dumas describes the sky as coloured in "unhealthy, yellowish tones" drawing an implicit parallel with the inhabitants of the Marshes, who also were usually described as having a yellowish and unhealthy complexion. This aspect is illustrated by an anonymous traveller, who visited the Marshes in the period between 1821 and 1822:

> Of the Pontine Marshes, the first scenes our opening eyes beheld, little confirmation is required to prove the deadly effects of their exhalations; the livid aspect of the peasantry tell the tale but too forcibly; and the best comparison I can make as to their appearance is, that their ghastly complexion resembles the yellow orange that blooms around. The vapours which affect their health seem also, though naturally, to affect their powers of action, for more wretched habitations, greater filth, and more squalid misery, I have nowhere beheld.
>
> (Anon. 1824: 45–46)

This traveller makes explicit the relations between atmosphere, landscape and the human body: the Marshes' vapours were thought to penetrate into human bodies and to have an influence on the region's inhabitants. It seems like if breathing the particular air of the Marshes was seen as tantamount to becoming part of it, to acquire its features. Indeed, in the 18th and 19th centuries, as mentioned above, the air was thought of as fluid, dependent on geographical and meteorological conditions, able to affect the living body in multiple ways (see Corbin 1986: 11).

It is for this reason that travellers to the Marshes, as well as physicians and experts of public health all over the world, were interested in understanding the material qualities of the air, in order to detect the potential "dangers in the atmosphere" (ibid.:7). The French historian Alain Corbin, for instance, when inquiring into the perception of disease and miasmas in the 19th century focusing on the sense of smell, has argued that:

> the sense of smell locates hidden dangers in the atmosphere. Its capacity to test the properties of air is unmatched. . . . The nose anticipates dangers; it recognizes from a distance both harmful mold and the presence of miasmas. It is repelled by what is in a state of decomposition. Increased recognition of the importance of the air led to increased acknowledgment of the importance of the sense of smell as an instrument of vigilance.
>
> (Corbin 1986: 7)

However, the materiality of atmosphere and its dangers were not only perceived through smell. For example, travellers in the Pontine Marshes detected these dangers in the visual and aural aspects of atmosphere, rather than in its odour. This aspect is emphasised in many travel diaries such as the one by Edward Daniel Clarke, a British traveller who describing his journeys in the Pontine Marshes and broadly in the Mediterranean region, writes:

> Places infected by such dangerous vapour may be distinguished, at the setting or rising of the sun, by thick and heavy mists of a milky hue; these may at that time be observed, hovering and rising high above the soil. [. . .] In those countries, swarms of venomous insects, by the torments they inflict, warn mankind to avoid the deadly atmosphere. . . . The noise made by them is louder than can be imagined; and when joined to the clamorous whooping of millions of toads (such as the inhabitants of northern countries are happy never to have heard), silence the ordinary characteristic of solitude is completely annihilated.
>
> (Clarke 1812: 652–653)

Clarke points out that the dangerous atmosphere of the Marshes is recognisable from the presence of vapour and insects, and from the sound made by toads. The same point is stressed in the diary of John Strutt, another British traveller who in May 1841 approached the Pontine Marshes from the north, travelling along the Appian Way:

> We were warned of our approaching the Pontine Marshes by the cessation of tillage; to which pasturage succeeds, till within about two miles of Torre tre ponti, where the meadow degenerates into a marsh, vocal with frogs, whose ceaseless croakings and squabblings I hear, even as I write . . . it will shortly be almost deserted, as the mal'aria commences to-day, we are told. . . . It would seem that the period of commencement of this mal'aria is precisely fixed; an unpleasant consideration for a hypochondriac, or *malade imaginaire*, who might be arriving at the exact moment; his only resource would be to follow our example, according to the advice of our landlady, in swallowing as much pepper as our throats would bear, in order to fortify our stomachs against an attack from the enemy.
>
> (Strutt 1842: 8–9)

In these notes the British traveller uses the word enemy with regard to the *mal'aria* ('bad air') highlighting its concrete presence that, as he writes, has a fixed date of commencement, and it also seems to be perceived through particular visual and aural signs: the cessation of tillage and the "ceaseless croakings and squabblings" of frogs.[8]

From these excerpts, the atmosphere emerges as a material substance, an object of perception and interaction, something that could be heard and seen. The mysterious vapours of the Marshes, then, understood as beautiful and deadly at the same time, became for travellers one of the main element of interest of the Pontine wet landscape. It certainly proved the most visible and concrete manifestation of their atmosphere, in that these vapours were understood as able to seep into the pores of skin and to affect the mind and the body of human beings.

The atmosphere as material object of perception and interaction

As I have demonstrated, until the beginning of the 20th century the notion of 'bad air' was used to describe a specific kind of atmosphere, associated with marshlands and generally with wet landscapes, and a particular disease whose origin was uncertain, namely malaria. This atmosphere was perceived as rising from the muddy soil of the marshes and creeping into human bodies. This is a particular interpretation that recalls the idea of atmosphere conceptualised by philosopher Gernot Böhme, seeing atmosphere as the subject of inquiry of a new aesthetics. According to Böhme, atmosphere is "in between environmental qualities and human states" (Böhme 1993: 114); atmosphere *is* the link between "environmental qualities and states" (ibid.). However, Böhme's states seem different from material bodies, and his atmosphere appears as a discarnate entity (see also Ingold 2015). Nevertheless, like the 'bad air' in the Pontine Marshes, atmosphere conceptualised by Böhme is "a phenomenon or condition that transgresses boundaries, such as object and subject" (Bille et al. 2015: 32). The German philosopher highlights this point arguing that:

> Atmospheres are neither something objective, that is qualities possessed by things, and yet they are something thinglike, belonging to the thing in that things articulate their presence through qualities-conceived as ecstasies. Nor are atmospheres something subjective, for example, determinations of a psychic state. And yet they are subjectlike, belonging to the subjects in that they are sensed in bodily presence by human beings and this sensing is at the same time a bodily state of being of subjects in space.
>
> (Böhme 1993: 122)

Böhme's notion of atmosphere is interesting to us here because it challenges and undermines the dualism between separation and unity, outside and inside, form and process, emphasising the idea of ecstasies, that is the way in which the qualities of a thing – i.e. the atmosphere – "goes forth from itself" (Böhme 1993: 121). Tim Ingold offers a clarification of Böhme's argument arguing: "[. . .] it is from the coming together of persons and things that atmospheres arise: they are not objective yet they inhere in the qualities of things; they are not subjective yet they belong to sensing beings" (2015: 77).

This idea resonates with Benjamin's concept of aura, which Böhme describes as "something which flows forth spatially, almost like a breath or a haze – precisely an atmosphere. Benjamin says that one 'breathes' the aura. This breathing means that it is absorbed bodily [. . .] that one allows this atmosphere to permeate the self" (Böhme1993: 117). While Böhme mentions the idea of breathing, it is worth mentioning that this kind of atmosphere that he is talking about is completely divorced from the element of air. The atmosphere of the Pontine Marshes, in contrast, was understood in a meteorological and material sense – as dependent on temperature, air, weather, wind and the condition of the soil. At the same time, the Marshes' atmosphere was also considered as an outcome of human activities, of their effort of 'cultivating the bogs' and maintaining the land wet. In short, this kind of atmosphere was *both* meteorological and affective.

The atmosphere of the Pontine Marshes, as interpreted in the 18th and 19th centuries, then, becomes interesting because it affords the possibility of conceptualising the notion of atmosphere beyond the boundary between meteorology and aesthetics. It represents a transgression of differences such as "nature and humanity, materiality and sensoriality, the cosmic and the affective" (Ingold 2015: 76). In this sense, my discussion of atmosphere follows the line traced by Ingold in conceptualising a holistic notion of atmosphere based upon a reassessment of the element of air (ibid.: 78). Nevertheless, the idea of air that emerges from the materials I have presented challenges the idea expressed by Tim Ingold who, in seeing the air as a medium, states:

> On no account, however, can the air be converted into an object that the child, or anyone else can have a relationship with. Thus the walker does not interact with the air as he sets his face to the breeze, but feels it as an all-enveloping infusion which steeps his entire being. It is not so much what he perceives as what he perceives in.
>
> (Ingold 2015: 70)

Even though the 'bad air' was certainly understood as "an-all enveloping infusion" for scientists and humanists who crossed or approached the Marshes, I argue that the atmosphere was more than a medium. It was something tangible which manifested itself through different phenomena that could be perceived through sight, smell and hearing. Moreover, as we have seen, the Marshes' atmosphere was considered as an enemy, beautiful and dreadful at the same time, as an object of aesthetic contemplation and scientific measurement, and as something against which to take precautions.

Indeed, even if the origin of the 'bad air' was uncertain, it was not perceived as vague as often highlighted as a characteristic of atmosphere (Anderson 2009: 78); it had a precise location (see Wilson 1820: 662), and a fixed date of commencement (see Strutt 1842: 8). The 'bad air' was understood as a concrete presence with which travellers had material relationships. In this sense, their experience of atmosphere was truly aesthetic "in the Greek sense of aistesis – 'sense experience'" (Anderson 2009: 79): a sensorial experience of atmosphere, which rather than being a medium of perception, was encountered as a tangible, material entity – both threatening and mesmerising at the same time.

Conclusion: bodies, landscape and atmosphere

As I have demonstrated, until the beginning of the 20th century the Marshes and their atmosphere were understood by scientists and aestheticians as part and parcel of a particular taskscape characterised by a specific array of agricultural activities. Upon reflection, it seems as if the Marshes and their atmosphere, in fact, were revealing a particular kind of human-environment relations grounded in the practice of cultivating the bogs. As means of concluding this chapter, I retrace my steps in the following, reflecting on my encounter with atmosphere in the Agro Pontino of today and its relationship to the historical materials presented above.

The area I described at the beginning of the chapter, the *Gricilli* area, hosts three major lakes, two of them filled with sulphur water, and other small ponds, springs and channels. The *Gricilli* area is inhabited by a family of farmers who have lived and worked there since before the fascist reclamation, and the site is locally perceived as being symbolically, demographically and ecologically related to the vanished landscape of the Pontine Marshes (see Gruppuso 2016).

In the larger area at the foot of the hills, the practice of burn-beating is widespread and it occurs every year in the early spring. As I mentioned, this activity is an ancient practice that was widely practiced in the amphibious environment of the Marshes. Nowadays this practice has decreased, and has also changed its goal: nobody burns the land in order to create ponds like in the Pontine Marshes. Burn-beating is now mainly related with aesthetic concerns, namely the idea of keeping the land clean, or economic concerns related to provide new and fresh pastures for animals, mainly sheep. Nevertheless, like the vapours in the old Marshes, which were an outcome of the local people's engagement in that particular environment, nowadays the smoky vapours coming from burning reeds are, even though for other reasons, the manifestation of a form of nurture and management of that wet landscape.

Speaking with the farmers who inhabit the area, it also appeared that, in their understanding, the origin of the lakes is deeply related to this particular agricultural practice. They explained to me that the lakes are situated in a particular area covered by a thick layer of peat which burned and was consumed little by little to an extent deep enough to be filled by the surrounding water. In this reading, the lakes are the results of the concentration of this water in areas where the ground sank due to the activity of burn-beating in combination with the quality of the soil. This explanation is interesting and pertinent to my argument because it demonstrates that if landscape is literally "a land shaped" (Ingold 2012: 198) by particular activities and hence, as I argued, a taskscape, then the atmosphere is actually part and parcel of that particular taskscape: it is air shaped by particular activities. Like the 'bad air' experienced by scientists and humanists in the old Marshes, the atmosphere which I experienced in the Gricilli area has also to do with the shapes of the land; it emerges along the practices of care which people perform towards the landscape in their process of inhabiting the world. In this sense, the smoky vapours which affected my emotional and sensorial sphere are reminiscent of the old accounts of the evanescent, vaporous, cloudy and misty nature of the Marshes' *mal'aria*.

To conclude, let me return to the notion of vapour in order to reflect on the indissoluble binding between body, landscape and atmosphere. Vapour recalls air, wind, water, materials and the relations amongst them. Vapour also recalls the physical and chemical reactions that, through heating, transform materials in aerial substances rising from the ground to the sky. The mysterious vapour in the old Marshes, as the vacuous columns of smoke in the *Gricilli* area, are actually phenomena that connect the earth to the sky. This point resonates with the argument that Clark and Yusoff made about fire. They argue that fire is "a force that binds intimate and mundane human activities to some of the most 'monstrous' energetic movements of the Earth. As Gaston Bachelard puts it, fire 'links the small to the great, the hearth to the volcano, the life of a log to the life of a world' (1987: 16)" (Clark and Yusoff 2014: 206).

This circular connection resembles the one which occurs through the process of breathing. As I have demonstrated in this chapter, to breathe the vapours coming from the Marshes was understood as tantamount to breathing their atmosphere and becoming part of it. Taking seriously the etymology of the word atmosphere as vapour in the sphere, I argue that as earthly beings, to engage with these vapours means to participate in the landscape, to reconnect the earth with the sky, the ground with the air. It means to bring the atmosphere within the realm of perception and activity. It means highlighting the ambiguous nature of the concept of atmosphere, and to connect its meteorological and aesthetic qualities with the terrestrial experience of living beings on earth.

Notes

1 I intentionally use the expression 'wet lands', instead of the much more common term 'Wetlands', to mean a holistic understanding of these environments independent from their political and ecological designations as protected areas (see Gruppuso 2016).
2 I carried out ethnographic fieldwork between spring 2011 and autumn 2012 as part of my Ph.D. research, at the University of Aberdeen. In my doctoral research I examined environmental conflicts between agriculture and nature conservation in two protected wetlands of Agro Pontino, Italy (see Gruppuso 2016).
3 Translation from the Italian by the author.
4 Translation from the Italian by the author.
5 Interview conducted 15 March 2011. Translation from the Italian by the author.
6 Plasmodium is the parasite provoking the infection called malaria.
7 Translation from the Italian by the author.
8 A similar understanding emerges from other travel diaries, which also strongly emphasise the materiality of atmosphere in relation to landscape and its inhabitants (e.g. De Forbin 1819: 64; Anon. 1819: 424).

References

About, E. 1861. *Rome of Today*. New York: James O. Noyes, Publisher.

Anderson, B. 2009. Affective Atmospheres. *Emotion, Space and Society* 2, 77–81.

Anon. 1819. Art XVII. Voyage dans le Levant, en 1817 et 1818. Par le Comte de Forbin. *The British Review and London Critical Journal* XIV(XXVIII), 405–438.

Anon. 1824. *Mementoes, Historical and Classical of a Tour Through Part of France, Switzerland and Italy, in the Years 1821 and 1822*, Vol. II. London: Baldwin, Cradock and Joy.

Bachelard, G. 1987. The Psychoanalysis of Fire. London: Quartet.

Bille, M., Bjerregaard, P. and Sørensen, T.F. 2015. Staging Atmospheres: Materiality, Culture and the Texture of the in-Between. *Emotion, Space and Society* 15, 31–38.

Blodget, L. 1857. *Climatology of the United States*. Philadelphia, PA: J.B. Lippincott and Co. London: Trubner and Co.

Böhme, G. 1993. Atmosphere as the Fundamental Concept of a New Aesthetics. *Thesis Eleven* 36, 113–126.

Breda, N. 2001. *Palù. Inquieti paesaggi tra natura e cultura*. Verona: CIERRE edizioni.

Cavallo, F. L. 2011. *Terra, Acque, Macchine. Geografie della bonifica in Italia tra Ottocento e Novecento*. Reggio Emilia: Diabasis.

Clark, N. and Yusoff, K. 2014. Combustion and Society: A Fire-Centred History of Energy Use. *Theory Culture & Society* 31(5), 203–226.

Clarke, E. D. 1812. Travels in Various Countries of Europe, Asia and Africa. By Edward Daniel Clarke. Part the Second. Greece, Egypt and the Holy Land, Section the First. *Supplementary Number to the Thirty-Third Volume of the Monthly Magazine* 33(229), 635–661.

Corbin, A. 1986. *The Foul and the Fragrant: Odor and the French Social Imagination*. Leamington Spa, Hamburg and New York: Berg.

Crotti, D. 2005. La Malaria, ossia la mal'aria: brevi note di una 'storia sociale e popolare'. A history of Malaria From and for a Popular Point of View. *Le Infezioni in Medicina* 4, 265–270.

De Forbin, L. 1819. *Voyage dans Le Levant*. Paris: De L'Imprimerie Royale.

Di Tucci, P. 1882. *Commissione per lo studio dei Boschi nella Provincia di Roma*. [manuscript] Archivio Centrale dello Stato. Ministero agricoltura industria commercio. Direzione.Generale. Agricoltura. 1° Vers., busta 526, fasc. 1463: VERBALI DELLE VISITE ALLE PALUDI PONTINE.

Dumas, A. [1869] 2007. *The Last Cavalier: Being the Adventures of Count Sainte-Hermine in the Age of Napoleone*. New York: Pegasus Book.

Edensor, T. 2010. Aurora Landscape: Affective Atmosphere of Light and Dark. In *Conversation With Landscape*, edited by K. Benediktsson and K. A. Lund. Farnham: Ashgate, 227–240.

Folchi, A. 1996. *L'agro Pontino 1900–1934*. Roma: Regione Lazio.

Folchi, A. 2002. *Le Paludi Pontine nel Settecento*. Formia: D'Arco Edizioni.

Goethe, W. J. 1817. *Goethe's Travels in Italy: Together With His Second Residence in Rome and Fragments on Italy*. London: George Bell and Sons, York Street, Covent Garden.

Gruppuso, P. 2014. *Nell'Africa tenebrosa alle porte di Roma: Viaggio Nelle Paludi Pontine e nel loro immaginario*. Roma: Annales edizioni.

Gruppuso, P. 2016. *From Marshes to Reclamation: There and Back Again. Contested Nature, Memories and Practices in Two Wetlands of Agro Pontino, Italy*. PhD Thesis, University of Aberdeen, Aberdeen.

Harris, L. E. 1957. *Land Drainage and Reclamation*. In *A History of Technology*, edited by J. C. Singer. London and New York: Oxford University Press, 300–323.

Hippocrates. 1978. *Hippocratic Writings*. Harmondsworth: Penguin Books.

Hutchinson, G. 1848. *Treatise on the Causes and Principles of Metereological Phenomena*. Glasgow: Archibald Fullarton and Co.

Ingold, T. 1993. The Temporality of the Landscape. *World Archaeology* 2(25), 152–174.

Ingold, T. 2005. The Eye of the Storm: Visual Perception and Weather. *Visual Studies* 20(2), 97–104.

Ingold, T. 2006. Rethinking the Animate, Re-Animating Thought. *Ethnos: Journal of Anthropology* 71(1), 9–20.

Ingold, T. 2007. *Lines: A Brief History*. London: Routledge.

Ingold, T. 2011. *Being Alive: Essays on Movement, Knowledge and Description*. London: Routledge.

Ingold, T. 2012. The Shape of the Land. In *Landscape Beyond Land: Routes, Aesthetics, Narratives*, edited by A. Arnason, N. Ellison, J. Vergunst and A. Whitehouse. New York and Oxford: Berghahn Books, 197–208.

Ingold, T. 2013. Making and Growing: An Introduction. In *Making and Growing: Anthropological Studies of Organisms and Artefacts* edited by E. Hallam and T. Ingold. Farnham: Ashgate, 1–24.

Ingold, T. 2015. *The Life of Lines*. Abingdon: Routledge.

Mammuccari, R. and Langella, R. 1999. *I pittori della Mal'aria dalla Campagna romana alle Paludi pontine*. Roma: Newton & Compton.

McCormack, D. 2008. Engineering Affective Atmospheres on the Moving Geographies of the 1897 Andrée Expedition. *Cultural Geographies* 15(4), 413–430.

Nicholas, G. P. 1998. Wetlands and Hunter-Gatherers: A Global Perspective. *Current Anthropology* 5(39), 720–731.

Nicholas, G. P. 2013. Towards an Anthropology of Wetland Archeology: Hunther-Gatherers and Wetlands in Theory and Practice. In *The Oxford Handbook of Wetland Archeology*, edited by F. Menotti and A. O'Sullivan. Oxford: Oxford University Press, 761–778.

Nicolai, N. M. 1800. *De' bonificamenti delle terre pontine*. Roma: Stamperia Pagliarini.

Ogden, L. 2011. *Swamplife: People, Gators, and Mangroves entangled in the Everglades*. Minneapolis, MN: University of Minnesota Press.

Secchi, A. 1865. *Sulle condizioni igieniche del clima di Roma*. Roma: Tipografia delle Belle Arti.

Strutt, A. J. 1842. *A Pedestrian Tour in Calabria e Sicily*. London: T.C. Newby.

Taussig, M. 2004. *My Cocaine Museum*. Chicago: The University of Chicago Press.

Tedeschini, M. 2014. Introduction. *Lebenswelt. Aesthetics and Philosophy of Experience* 4, 134–137.

Ten Bos, R. 2009. Towards an Amphibious Anthropology: Water and Peter Sloterdijk. *Environment and Planning D: Society and Space* 27(1), 73–86.

Traina, G. 1988. *Paludi e Bonifiche del Mondo Antico: Saggio di archeologia geografica*. Roma: L'Erma di Bretschneider.

Wilson, J. 1820. *A Journal of Two Successive Tours Upon the Continent, in the Years 1816, 1817, & 1818*. London and Edinburgh: T. Cadel & W. Davies.

5 Senses of being

The atmospheres of listening to birds in Britain, Australia and New Zealand

Andrew Whitehouse

Sense of place, sense of being

When the Listening to Birds project was launched in November 2007 I received several hundred emails from people describing their experiences of bird sounds.[1] One came from Lou Horton from Devon, who wrote:

> Birdsong becomes so much a part of the aural environment it becomes nearly invisible – until it changes. I came to the UK as a teenager having grown up in Australia. Two things struck me straight away: both the stars and the birds were wrong. More than anything else, these two things made me feel alien.
>
> Nearly 30 years later I came across Australian birdsong on the internet. A short burst of currawong song brought back an intense feeling of being a child again in Sydney. I could almost smell the air and feel the texture of my primary school uniform. It's like a trigger to a sense of being, rather than a memory of doing.

Lou's account reveals that, although birdsong is a feature of experience that many people rarely notice, it can still be very significant, something that a change in circumstances can suddenly reveal. Listening to birds can be subtly integral to the atmosphere of a place and to a sense of belonging, but can also be profoundly strange and alienating. It can also be a way to feel connections with other places and other times, as well as with the here and now. My aim here is to explore how my respondents understood their encounters with birds through sound, the atmospheres these were part of and the ways that these changes are involved in both their sense of place and, as Lou puts it, their sense of being. I develop further on these themes by considering how senses of place and of being relate to the more-than-human atmospheres that are experienced through listening to birds. To do this I consider the relations between senses of place, of being and of atmosphere. I examine how these experiences are narrated and how people's sense of being is related to shifts in the more-than-human atmospheres of listening to birds.

When the Listening to Birds project was launched, it was featured prominently in the media, most usefully on the BBC News website, which provided a link to the project website. This included a page through which people were invited to describe their experiences of bird sounds. Respondents were, perhaps inevitably,

English speaking. Most responses came from the UK but many were from and about Australia and New Zealand. I do not wish to suggest that listening to birds is as important to British people in general as it is to my respondents. Instead, I argue that the narratives I discuss are highly typical; they are an exemplification of how experiences of listening to birds can be narrated and the role this plays in sensing place and in the atmospheres that many people encounter in their everyday lives.

The responses varied greatly in length and subject matter, some being just short comments and others being detailed reminiscences or descriptions of how bird sounds are recognised. Although almost all described bird sounds very positively, few claimed to be birders; these were not people who go out with the specific intention of encountering birds. Instead, most encountered birds primarily whilst doing something unrelated, such as lying in bed, sitting in their garden, or travelling to work. In most cases people wrote about listening to birds that live alongside respondents and that were, at least at some point in their lives, everyday encounters. Blackbirds, for example, are significant to many in Britain in part because they live around people, in gardens, parks and woods. Their lives are entangled with many and this presence is revealed most readily and most eloquently through their singing and vocalising.

I understand the encounters I describe as respondents' narrations of being alongside nonhumans in a shared world (Whitehouse 2017). In this sense, what they write is their way of understanding their experience at that specific point but that draws upon the narrator's accumulated life experience (Rapport and Overing 2000: 285). My interest is in how these narratives of experience are distinctly personal in reflecting the sometimes powerful yet everyday experience of listening to birds, but I will also touch on how these experiences are influenced by larger historical processes, narratives and ideas. In particular, I examine how bird sounds provide a focal point that is ecologically bound with various other elements in people's experience of atmosphere. I also discuss how certain aesthetic and moral sensibilities are brought to bear on these experiences and influence the way that atmospheres are reflected upon and felt.

I explore these themes in three stages. First, I consider the homely atmospheres that emerge through the sounds of familiar species. I also address the question of how bird sounds come to be noticed in the first place and how everyday experiences of listening to birds become aesthetic. I argue that aesthetic understandings of bird sounds need to reflect the whole bodily experience of listening rather than just treating the sound as an isolable aesthetic object. The aesthetics of the more-than-human atmospheres that emerge through listening to birds thus focus on how diverse elements are gathered together in being sensed, felt and narrated. Second, I discuss the ways in which changing circumstances can reveal how experiences are idealised and related to larger narratives of nature and nation. Changes, for example in the birds that people can hear, reveal an experiential aesthetics of place in which the atmospheres that are felt can seem homely or unsettling. Third, I look at examples of listening to birds in New Zealand where the more-than-human atmospheres elicited draw together places and senses of being but also evoke more unsettling biographical or national experiences in which atmospheres can be felt as 'out of place'.

Before discussing these themes, I will explain my approach to senses of place and of being and how I see these as relating to the more-than-human atmospheres of listening to birds.[2]

Sounds, charisma, place and atmospheres

Places are sensed. Here, my focus is on how places are sensed through listening to birds but I first emphasise that listening is a multi-sensory experience that has sound as a focal point that gathers together a range of other elements. It is this gathering that respondents describe in their narratives. It is also this gathering of elements that can be defined as an atmosphere. Atmospheres are thus diverse mixtures that are inherently more-than-human (Ingold 2015: 72).

Places are also ongoing. Massey has argued for "an understanding of both place and landscape as *events*, as happenings, as moments that will again be dispersed" (2006: 46 emphasis in original). This dynamic sense of ongoingness is apposite to how places are experienced through bird sounds and the atmospheres that cohere around them. By their very nature these arise, with regularity but not certainty, both *over* time and at *specific* times. The work of Feld is particularly relevant here, not only for his work on the poetics of bird sounds in New Guinea (1990) but in his concept of acoustemology: the way that places are sensed through sound (1996, 2000). My approach here, however, is broader; whilst sound is the focal point, I argue that the contextualisation of sound and the ways it becomes meaningful are normally perceived through a range of other sensory modalities. Sound is thus not isolated as 'the soundscape' (Ingold 2011: 136–139) but is focal to a whole bodily experience. The approach I put forward is not simply acoustemological but is instead an investigation of how the atmospheres of a place are narrated. It is through this multi-sensory experience and process of narration that places come to be felt and understood.

Atmospheres occur in places. Experiencing an atmosphere is an aspect of sensing place, but not exactly the same thing. The distinction is necessarily imprecise and to draw a sharp line between the two would be to misunderstand their relationship. Places are inhabited; they are where life happens and through living they come to be known by their inhabitants. Places *have* atmospheres. These atmospheres, though recurring in some respects, are more ephemeral than the places where they occur. They come and go, shifting sometimes abruptly as weather, activities and light change. As such, part of the distinction between a sense of place and of atmosphere is the temporal duration of what is being experienced. Atmosphere is a way of referring to the ongoingness of places, to those changeable gatherings that are bound up with being somewhere. They emerge through situated relations and need to be understood both in terms of the ecologies of the phenomena encountered and the perceptual and meaning-making practices of those experiencing them.

Atmospheres are neither subjective nor objective (Anderson 2009). They are experienced as external to the self but are also felt in ways that are personal. The gatherings that make atmospheres are not simply generated by a singular observer but are brought together by the myriad mixtures and goings-on of places.

Atmospheres are more-than-human and more-than-subjective. They are always, however, felt in ways that may be shared but are also personal. Atmospheres draw together environment and sentiment (Ingold 2015: 79), the cosmic and the affective. Atmospheres emerge through the activities of beings that are inherently perceptual and, thus in turn, the gathering of atmospheres is responsive to what is going on. What is noticed and most strongly felt is personal but is often shared (Anderson 2009: 80). Often there is a focus of activity that draws attention to certain aspects of what is gathered and this also elicits certain feelings. Sounds are atmospheric elements *par excellence*, breathing life into lines of flight and movement (Ingold 2015: 111). Even if they are not a focal point, sounds provide texture to the experience of places. This background texturing can become more apparent in changing circumstances when different sounds are heard, revealing contrasting life lines and mixtures.

The feelings that are elicited through atmospheres are integral to a sense of being. Lou made a distinction between memories of doing and triggers to a sense of being, and this provides a basis for understanding the kinds of memories people have of different times and places and, through those memories, understanding change. I asked Lou to clarify what she had meant by the distinction she drew, and she answered by saying that, for her, memories of doing were quite specific events and activities that were provoked at certain times. A sense of being is less specific but more visceral and more powerful. When she heard the recording of a currawong it suddenly triggered a sense of what it was like to be a child again in Sydney, rather than any specific thing she did as a child. Hearing a sound elicited a sense of how it once gathered together other elements that recurred in her childhood, such as the feeling of her school uniform and the smell of the air. This atmosphere was made fleetingly tangible by the recording and, in turn, an analogous shift in her sense of being emerged.

A sense of being is bound up in feelings that in turn draw upon a sense of what is moral and aesthetic about experiences. The atmosphere of a place is thus implicated in senses of, for example, homeliness or of being alien, as Lou mentions. These feelings are prominent in narratives of listening to birds. In turn, they are influenced by larger narratives that extend the specific experiences outwards. These can include biographical or national narratives, for example. All of these elements can mix together to provide a further sense of whether a place, and one's sense of being in it, is as it should be.

A sense of being is aesthetic and affective. In the case of listening to birds, the senses of being evoked are bound up with relations with nonhuman animals. Two geographers have considered these dimensions of human-animal and human-bird relations, and I take their approaches as a point of articulation for my own.

Ornithophilia and nonhuman charisma

Mark Bonta (2003, 2010) draws on Wilson's concept of biophilia to develop 'ornithophilia' as a means of exploring how and why certain people form an affinity with birds. Jamie Lorimer (2007), meanwhile, has theorised the dimensions of charismatic species: why certain kinds of animal seem to hold a strong appeal and how this appeal is then utilised by conservationists. On the face of it, these two authors are approaching affective human-animal relations from opposite directions: Bonta is drawing attention to the ornithophiliac person whilst Lorimer is considering the charismatic qualities of particular species. To an extent, the apparent opposition in the two approaches is a product of the different foci of each author; Bonta considers birdwatching as an activity that involves people who engage in ongoing relations with place whilst Lorimer explores the political usage of charismatic 'flagship' species within conservation. Though their emphases differ, both accounts draw humans and animals together in an experiential process and event.

Bonta's investigation of ornithophilia emphasises birding as an attunement of the senses that layers significance into encounters entangling birders with the avian landscape. Those who become 'infected' with ornithophilia (2010: 142) are thus drawn into the temporal rhythms of birds and the places they inhabit. But despite this entanglement of people, birds and landscape, Bonta argues that "The aesthetics in birding involves the experience of beauty (the sublime) *inherent* in the colours, flight, songs, calls, and other, less often noted characteristics of the avifauna" (ibid: 149 my emphasis). From Bonta's own account there is clearly more to the aesthetics of birding as an activity than this, although birders are rather adept at perceiving birds as objects with inherent qualities and can sometimes, as Bonta points out, be indifferent to the sometimes unseemly locations where they go birding. My own interest is less on the highly focussed engagement of birders but on narratives of listening to birds in more everyday settings. It concerns how bird sounds become important to the experience of atmosphere and how these experiences are reflected upon. As such the aesthetics that are invoked are *emergent* from the relations of encounter rather than inherent in objects of observation.

Turning to Lorimer, he defines nonhuman charisma as, "the distinguishing properties of a nonhuman entity or process that determine its perception by humans and its subsequent evaluation" (2007: 915). Charisma is thus relational; it is not so much inherent in a nonhuman as emergent in its perception by humans, with their own sensory particularities. Where Lorimer's approach is lacking is that he focuses on charisma as emergent in the human perception of the nonhuman animal *in itself*. What this leaves out is the *situatedness* of the encounter and the bearing this has on its 'subsequent evaluation' by humans. He ignores the atmospheric gathering together of elements that, for example, a bird sound is embedded within and, in turn, draws attention to. Instead, the affective taxonomy that Lorimer outlines is focused on the bodily qualities of nonhumans and on the perceptual capacities of humans to discern those qualities, particularly visually, and their subsequent emotional responses.

This, I argue, is too narrowly-focused an approach to understand the appeal of engagement with nonhumans and the places they are bound up with. Indeed, Lorimer introduces his article by discussing Craig, a conservationist who surveys corncrakes in the Hebrides. What Craig finds appealing is not so much 'the corncrake' or even its rasping and repetitive call, but "the crake-filled summer nights out on the islands" (2007: 911). It is thus the situated and broadly-encompassing atmosphere that is brought into focus by the sound of the corncrake that is charismatic and not simply a more narrowly-defined engagement of human and bird bodies. My approach to the feelings and aesthetics described in the narratives below is that they are elicited through their incorporation into ongoing atmospheres that listeners experience as they go about their lives. They are as much about the experience of a shared world as they are about encounters with specific other beings.

Blackbirds, home and noticing bird sounds

Whilst I have never calculated which bird is most frequently mentioned by respondents, I suspect that, were I to do so, it would be the blackbird. Here is a typical account:

> [I love the song of the] blackbird especially at dusk on a sultry evening after a hot sunny day. When everything is still. No wind. Very warm. About 10 p.m. at night, still light and reminds me of being little again.
>
> (Barbara Adams, Darlington)

In this account, and numerous others, the mellifluous, relaxed song of the blackbird is associated with certain conditions and times of day. The warmth and calmness of the blackbird song becomes focal to the sultry atmosphere of spring and summer evenings, mixing with the scent of garden flowers and the light gradually fading into dusk. It is almost as if the song gives off its own kind of warmth. As one respondent succinctly puts it:

> Blackbird – long warm evenings. Time to get the deckchair out.

Blackbirds also make other evocative sounds, particularly the insistent pinking call given at dusk as they congregate for their wintertime roost. Though less musical or beautiful than the song, this sound was often described as characteristic of a wintery atmosphere, just as the short winter days move into the cold darkness of evening. Blackbirds, by various means, are adept at making themselves heard and at drawing attention to their own place-making activities and in doing so they make many British people feel at home too. Indeed, some respondents even suggested that the blackbird's song is quintessentially British. But what is it about hearing a blackbird that makes people feel this way? To begin to address this question, I introduce some general points about listening to birds and the characteristics of the narratives I discuss.

The first general point is that bird sounds are important to people because of their very presence. Birds live amongst people in and around homes and gardens, in city centres and parks and in any number of rural locations. They are usually present throughout the year but, while this means that birds can be encountered in an everyday sense, their presence is not static and constant but changes with time and season (Whitehouse 2017). The dynamic presence of birds and the ways this presence entangles with people's lives, the changes, constancies and rhythms within it, is essential to making them significant and integral to ongoing atmospheric mixtures. They are important because they are there.

But how is this presence revealed? Of course one can see birds, sometimes easily, but many take more of an effort to see and to do so clearly and in a way that enables identification by sight often requires the aid of a pair of binoculars. Hearing, on the other hand, only requires that one listens. The ease with which the presence and identity of birds is revealed through sound is crucial to their significance in many people's lives. This means that listening to birds can readily be incorporated into many activities that take place within earshot of their sound-making. People hear birds when they are at home, in their garden, travelling to work and in bed at night. This is not to dismiss the enormous significance of seeing birds, but hearing birds is often one of the most straightforward and prevalent ways in which nonhuman animals come to be perceived. The atmospheres of bird sounds are ubiquitous and can even transcend the boundary between outdoors and indoors.

The prominence of listening and, most importantly, the atmospheric qualities and *situatedness* of bird sounds are highly significant to many respondents. Though this might seem obvious, it is something frequently overlooked in approaches that attempt to understand the aesthetics of bird sounds. In 1973 the philosopher and ornithologist Charles Hartshorne wrote *Born to sing: an interpretation and world survey of bird song*. The aim of this book was to apply musicological principles to bird song and to explore, as Hartshorne put it, "the possible scientific uses of the aesthetic analogy between . . . birds, and man with respect to music" (1973: 4). Hartshorne's survey was an ambitious attempt to reach beyond the subjective and survey the deeper musical structures of bird song. The songs of a huge range of species were systematically graded using various structural aspects such as repetitiveness and variety, so that in the most beautiful songs an optimum level of complexity is attained (1973: 8–9).

For Hartshorne, however, the beauty of the song must be assessed in isolation from the experience of listening to it; it needs to be severed from the atmospheric mixtures through which it emerges. Through disentangling oneself from the circumstances of listening, the sound in and of itself can be systematically analysed in comparison to others. The gathering brought into focus by sound is shorn away through analysis to reveal an object but not an atmosphere. This is not the way that bird sounds are really encountered or why they become evocative, affecting or beautiful to people.

Hartshorne is not alone in this objectifying analysis of song. Most scientific writing about bird sounds attempts to objectify the sound-making into 'the song' or 'the call' rather than to conceptualise bird sounds as ongoing threads of communication occurring concurrently with other activities and processes. This rather specialised form of classificatory analysis seems far removed from how most people, and very probably birds themselves, listen to sounds and seems unhelpful when considering the sorts of narratives discussed herein. In fact, although respondents often describe the sounds of birds as aesthetically pleasing as sounds, they sometimes find sounds that they describe as unpleasant to be equally evocative, meaningful and atmospheric. It is not a systematic, or even unconscious, judgement of the beauty of the song that influences the sorts of bird sounds that people find meaningful or moving. Instead, it is the atmospheric circumstances of listening to birds with which people share their lives and world that is pre-eminent; the 'beauty' of listening to birds emerges through its ongoing experiencing and not from contemplating an isolated object.

These narratives, I suggest, can be more fully reflected and understood through a focus on an aesthetics that is perceptual, situated and experiential and that takes as its first principle the co-presence of humans and birds in a shared world. This approach draws on Bateson's definition of aesthetics as "responsive to *the pattern that connects*" (1979: 8 emphasis in original). It follows from this that an aesthetic question is one that considers the connections between one organism and another and also the situatedness that gathers them into a shared, atmospheric world. It is these sorts of perceptual engagements and what people make of them that are central to an experiential aesthetics.

A further point arising from the narratives is that listening to birds is a whole bodily experience of place in which the sound is the focal point. Sounds are described together with smells, sights and the relative movement of air, as in this example:

> We used to live in Hampshire and had a large mimosa tree just outside our bedroom window. It would come into its glorious yellow bloom in March. At dawn, a robin habitually sang from this tree. The mimosa's sweet perfume and the robin's melodious song would drift through the bedroom window and one felt an overwhelming feeling of peace – that all was well in the world.
>
> (John Wolstenholme, London)

Here the senses are aesthetically drawn together in a memory of situated, atmospheric experience. The robin's singing is made more powerful by the colour and scent of the mimosa blooms, as well as the time of day and season. The activities of birds resonate (cf. Ingold 2000: 196; Wikan 1992: 463) with the rhythms of time and season to create an evocative atmosphere. Following from this understanding, many respondents expressed feelings of well-being in response to hearing birds and this seems most apparent when people's lives also resonate with these rhythms, a sense that comes from a sympathetic attention to the activities of other beings around us. Much as musicians in an orchestra attend to one another as they engage in their playing, people attend to birds within the current of their everyday activities. This attendance is bound up with atmosphere; a resonance between listener, bird and other aspects of their shared world evokes a sense of time and place but also other feelings, such as the peacefulness mentioned in the above example. Atmospheres gather together this mixture of elements in ways that evoke many different feelings, and listening to birds intensifies the experience more acutely. Listening to the commonplace, ongoing sounds of blackbirds and robins is not simply a case of hearing musical notes and tones; it is an experience that draws attention to the gathering of varied elements in a shared world that is both dynamic and recurrent.

Experiencing change in Britain and Australia

When I quoted my respondent Lou earlier she mentioned that a change in circumstances can reveal aspects of the everyday experience of place that were once barely noticed. This is particularly important because it demonstrates that normally peripheral experiences are not insignificant; not being conscious of some aspect of one's surroundings is not necessarily the same as indifference. Atmospheres lurk in the background of perception much of the time, but they still exert a powerful influence. Large-scale movement is not essential for people to notice changes in the birds they hear. Even over short periods of time, the landscape and birds in a single area can alter quite dramatically. Sometimes home can come to seem less homely through such changes. One respondent described how a new home needed to be found for him to feel at home:

> I moved to France from the UK three years ago and one of the benefits has been the rediscovery . . . of birds and their wonderful calls – unfortunately all too rare in an over-congested UK. . . . It's the call of the skylark that really brings back memories of very hot summer days as a young child wandering with friends across parched ploughed fields in East Anglia, the sun beating down and the incessant twittering from on high of the skylark and more often than not I could never find the bird in the sky. Today, to walk out of my own front door and hear the sound of the skylark as I work is without doubt one of my greatest pleasures.
>
> (Trevor Aylett, Abilly)

Here, learning to live in a new home provides a link with past experiences that are unattainable in a changed homeland. Many other respondents have moved between Britain and Australia, and their stories are often founded on stark differences in the sounds of birds, as this example illustrates:

> We have been here in Sydney, Australia for just over six months and soon discovered that, to the British ear, the Australian birdsong is really quite disruptive. We have heard of people emigrating *back* to the UK because of the 'ugly' birdsong here. In a nutshell I would describe the subconscious effect of 'birdsong' here as being to raise people's tension. It is a series of screeches or other worldly sounds. In the UK you wake to the blackbird, sparrow, or if you are lucky thrush, gentle, harmonious songs that usher in the day to come. Here the birds literally crash into your consciousness . . . I honestly believe that if you hooked somebody up and exposed them to British birdsong and then Sydney birdsong you would see the latter send the pulse racing.
>
> (Eugen Beer, Sydney)

This example is more extreme than most, and in some cases people moving from Britain to Australia have been excited by the new and exotic sounds they encounter. Here the perhaps strange idea of 'British' and 'Australian' birdsong is put forward as a contrast to which the 'British ear', attuned to the harmonious native sounds of home, is suddenly disrupted by the dissonance of Australian birds. These contrasts seem to emphasise an analogy between nation and fauna in ways that imply a kind of avian ethnicity. People may speak English, but the birds ensure that the two places sound and feel very different. It also emphasises the point that the atmosphere feels a certain way not just because of what is happening there but how it is experienced. Australia and its birds generate contrasting atmospheres if one is attuned to Britain and its birds. Perhaps over time this dissonance can be alleviated through re-attunement, as happened in Lou's experience.

The differences are not absolute though. There are blackbirds in Australia, and hearing them in a land to which they have been introduced provokes interesting responses:

> As an Australian living in London, bird song contributes strongly to my sense of place. I have recordings of some Australian birds in my iTunes collection that I listen to sometimes to remind me of home: cockatoos, whipbirds, currawongs and bellbirds are particularly evocative for me. . . . But I also have some British birds that I like in my collection, particularly the blackbird and the stonechat. I love the fact that the blackbird's call is often in the background in many different parts of the UK, so I associate it strongly with living here, and have gotten quite disoriented in Melbourne in Australia where blackbirds also live.
>
> (Adam Schembri, London)

19th-century settlers to the antipodes must have missed those sounds from the old country, and there were many attempts to establish birds from home and elsewhere. This means that some modern Australian cities have a cosmopolitan avifauna, as described in this account:

> When I was young, the birds I mostly heard around my home in Melbourne, Australia, were introduced species: common blackbirds, house sparrows, common mynahs, spotted turtle-doves and common starlings, but also native red wattlebirds and silvereyes. The sound of them has always brought back memories of those times. Now that I watch birds, the sound of those introduced pests annoys me. I can't hear the 'proper' birds over their din. Yet they still bring back fond memories.
>
> (Anonymous, Altona)

Here the sense that a place is not quite as should be emerges: the birds that are there are the wrong birds, in some cases. Species brought over by settlers at least in part to create an atmosphere of home that would aid in their own acclimatisation now make some native Australians feel less comfortable (Franklin 2006). Sometimes, an atmosphere evokes the guilt of colonialism and the ecological damage wrought by it. People's perception of atmospheres is not simply direct but is also influenced by broader narratives and ideas that shape how they are felt and known.

These narratives of life change and memory reveal idealisations of how everyday life should be experienced and sensed; they are thus about an aesthetics of experience. Changes through life may be adjusted to but the sounds that disappear or are lost can still be missed. But what seems most significant is that changes in the birds that people hear reveal something of what is personal and significant to their own lives and the worlds within which they emerge. Belonging to a place has long been understood as an interactive process (Barth 1969; Cohen 1987), but these narratives exemplify how the interactions through which people belong are with many aspects of the environment and emerge through varied sensory experiences. The more-than-human atmospheres that are noticed through changing circumstances shape the ways experience is judged and understood and the ways in which one's sense of being in a place is felt. Whether the atmosphere seems homely or alienating emerges *in-between* the ongoing, multi-sensory experience of the world and the personal or historical narratives these experiences recall and elicit (cf. Anderson 2009: 78; Bille 2015: 268).

Birds and home in New Zealand

The very different sounding avifaunas of Australia and Britain can generate atmospheres of both excitement and alienation for those that move between the two countries. But for those that travel from Britain to New Zealand, the experience is more complex still, and in reading the accounts of respondents one could be forgiven for thinking they are writing about entirely different places. There are some who are struck by how different the birds of New Zealand sound; others who are delighted to find the same birds that they knew from the homeland; some who are amazed at the profusion of bird song, whilst others who are perturbed by the beautiful but eerily silent native forests. It is no surprise that one encounters this variety of experiences of listening to birds in New Zealand when one considers the ecological isolation of the islands together with the relatively recent human history, in which the avifauna has become quite strikingly entangled.

Ecological histories of New Zealand, (e.g. Wilson 2004; Young 2004), posit three waves of extinctions since humans first visited around 2000 years ago, leaving *kiore* or Pacific rats behind, which predated on birds' eggs. The second wave began with Polynesian colonisation and the third with European arrival in the late 18th and 19th centuries. Europeans brought with them a wide range of species from their homeland. The temperate climate and the rapid Europeanisation of much of the landscape enabled many species, including songbirds such as blackbirds, song thrushes and skylarks, to establish themselves with great success. The highly distinctive native avifauna fared much less well, particularly in the wake of the arrival of ground predators, such as rats, stoats and possums. The New Zealand avifauna of the 21st century is thus distinctly heterogenous in origins, and any discussion of it, either by biologists or by my respondents, is bound up in talk of origins, of native, introduced and naturalised species. Violent, colonial narratives are routinely a part of the experience of more-than-human atmospheres in New Zealand.

The New Zealand of the late 18th century was a very different place to Britain though. Perhaps the first European to be struck by and to narrate their encounter with birds was Joseph Banks, the naturalist on James Cook's expeditions. In 1770 Cook's ship, *The Endeavour*, was anchored at Ship Cove in Queen Charlotte Sound, from where Banks wrote the following:

> This morn, I was awakened by the singing of the birds ashore from whence we are distant not a quarter of a mile, the numbers of them were certainly very great who seemed to strain their throat with emulation perhaps; their voices were certainly the most melodious wild music I have ever heard, almost imitating small bells but with the most tuneable silver sound imaginable to which maybe the distance was no small addition.

The atmosphere that Banks was sensing emerged through the sound of bellbirds, and it is still possible to hear a profusion of these around Ship Cove, and more particularly on nearby Motuara Island, a predator-free sanctuary managed by the New Zealand Department of Conservation. Since the acclimatisation societies introduced numerous species of bird, plant, fish and animal in the late 19th century, native birds have been in retreat and now some are only found on island sanctuaries such as Motuara and Tiritiri Matangi. This has meant that in many parts of New Zealand one mostly hears introduced European birds. Hearing native birds is a less-than-everyday experience for most 'Kiwis', who obviously relate to their native birdlife in conspicuous ways. Those that are heard most frequently, such as tuis, fantails and grey warblers, are those mentioned most often by New Zealand respondents. For those coming to live in New Zealand from the UK however, the introduced birds are comforting and the native birds are sometimes fascinating and sometimes strange. One respondent who moved from Scotland to Tauranga writes:

> It is very pleasing to me to hear the blackbirds singing in New Zealand. I am not aware of whether they are native or introduced, even so I love them. . . . There are also skylarks in the park close to our new home and they are reminiscent of spring/summer days back in what used to be a warm west of Scotland, flying up high and singing their little hearts out then plummeting to the ground again.

Here the sounds evoke an atmosphere that draws together different places and different biographical points in the respondent's life. The sense of being in Scotland and New Zealand is drawn closer, but sometimes this sort of connection is missing. New Zealanders in Britain write of missing the sound of tuis or fantails, whilst one respondent who moved from the English Midlands to Dunedin writes:

> I moved from Northamptonshire to New Zealand when I was eleven. There were many things I expected to miss when I came to New Zealand – friends, winter Christmases, familiar television programmes, and the like – but the

> most evocative single thing I have missed in the years since is the cawing of crows at twilight. The birds here in New Zealand have their own sounds – even species I know from Britain, like blackbirds, sound different here. . . . But there are no crows here, and the sound of crows still makes me homesick. There's one particular song, "Senses Working Overtime" by the band XTC, which ends in the sound of crows cawing. It always reminds me of childhood in a south Midlands village in the 1970s.

Here the atmosphere evoked by the recording is missed in a way that seems to reflect an uneasy sense of being in the new land. A narrative of movement draws attention to gaps in what is experienced that are related back to how it is assessed and felt. In Bille's terms there is a merging of 'the atmosphere that *is* and *should* be' (2015: 269 emphasis in original). In New Zealand a problem for some seems to be that other kinds of birds are missing from daily life: the native birds that have become restricted to island sanctuaries and remote areas of forest. Recently there have been a number of attempts to create 'island sanctuaries' on the mainland, of which the Zealandia sanctuary in central Wellington is perhaps best known. Surrounded by predator-proof fences, these bring the sounds of native birds into the city. Reintroduction campaigns are described in terms of 'bringing birds home', as if they were in exile on offshore islands, and gardeners are encouraged to plant native flowering shrubs to encourage the tui 'back home' and into their gardens. By bringing birds home and filling in the gaps in experience that are widely noticed, it seems they are also being incorporated in the process of fostering a homelier atmosphere for New Zealanders too.

Listening to birds and perceiving atmospheres

I began by raising the question of how hearing birds contributes to people's sense of place, or their place-making. My argument has implied that bird sounds are not just integral to a sense of place but to a sense of being, or that sense of being a particular person in a particular time and place. Who one is and where one is sometimes come to seem like the same experience (Bender 2001: 13). Birds, too, seem to be doing the same thing through their sound-making. They sing to establish relations with other birds and to make and mark out a home. They call together when they go to roost for the same reasons, and in doing all of this they come to people's attention as part of their own place-making. These processes of sound-making and place-making that in turn shape a sense of being are inherently atmospheric; they are ongoing mixtures of elements that are both physical and felt.

To humans, the sounds of birds are indexical of times and places. Birds are encouraged to sing by certain sorts of conditions, of light, weather, habitat and by other sounds that go on around them. Their sounds are understood to emerge because of time and place and are in themselves intended to further those relations. The experience of listening to birds thus represents certain kinds of relations, sometimes of difference and sometimes of continuity. It is these sorts of ongoing, atmospheric relations through which senses of both place and being emerge and through which people are situated. As Lou put it to me:

> The natural world is not a tableau of scenery – it is a living, rustling, humming, singing thing which we stand in the middle of.

This captures the experience of more-than-human atmospheres that emerge through listening to birds. Atmospheres are not easily reducible to imagination or feeling, to a sense of place and time, or to the purely external and material qualities of the world. All of those are bound up in atmospheres but atmospheres have a distinct ontological status from any of those phenomena. They are not simply imagined but emerge from a complex meshwork of interactions involving many living beings going about their lives (Ingold 2011). Atmospheres are thus also fluctuating and emergent; they are unfolding events rather than static objects (cf. Anderson 2009). They are never quite the same from moment to moment, even though regularities can still be sensed and related to memories of places and times. When birds are singing, their actions are bound up with the sounds of all those around them, with the acoustic conditions and with the growth and movements of other beings. Their singing is influenced by the prevailing seasons, weather conditions and changing light, as well as to other sounds in their environment. The atmospheres to which their singing contributes gather together all these elements.

As Lou points out, we stand in the middle of this atmospheric world. The atmosphere that we experience comes about not simply through what is external to us but also through our bodily capacities to perceive. An atmosphere in part is what we are sensitive to. Atmospheres are also inherently meaningful, but these meanings have certain specific qualities. Atmospheres are primarily felt, and this initial tacit and inarticulate feeling is essential to their semiotic power. Atmospheres can be reflected upon, as they are in the narratives discussed here, and this reflection can influence the ways they are subsequently sensed, but that reflection follows from those initial, often tantalising feelings. Atmospheres cannot entirely be captured in conscious reflection but those feelings can be evoked in the sort of 'Proustian rush' that Lou describes in her narrative at the beginning of this chapter and that others also convey. Atmospheric reality has a prominence and power that draws listeners out into the world by triggering intense but inarticulate feelings. This initial sensation is not the end of one's encounter but elicits further reflection on one's sense of being in the world and to the other beings within it.

Notes

1 This research was funded by a grant from the Arts and Humanities Research Council, number AH/E009573/1.

2 Focussing on the themes of birds, their sounds, senses of place and atmosphere aligns this paper with two areas of theoretical discussion: the sensing and experience of places in terms of their atmosphere and the emergence of relations between humans, animals and other nonhumans. Recent work in both geography (e.g. Cloke and Jones 2002; Massey 2006) and anthropology (e.g. Bender 2001; Feld 1996; Ingold 2000, 2011) has productively attempted to draw both areas together.

Bibliography

Anderson, B. 2009. Affective atmospheres. *Emotion, Space and Society* 2, 77–81.
Barth, F. (ed.) 1969. *Ethnic Groups and Boundaries: The Social Organisation of Culture Difference*. London: Allen & Unwin.
Bateson, G. 1979. *Mind and Nature: A Necessary Unity*. Cresskill: Hampton Press.
Bender, B. 2001. Introduction. In *Contested Landscapes: Movement, Exile and Place*, edited by B. Bender and M. Winer. Oxford: Berg.
Bille, M. 2015. Hazy Worlds: Atmospheric Ontologies in Denmark. *Anthropological Theory* 15(3), 257–274.
Bonta, M. 2003. *Seven Names for the Bellbird: Conservation Geography in Honduras*. College Station: Texas A & M University Press.
Bonta, M. 2010. Ornithophilia: Thoughts on Geography in Birding. *Geographical Review* 100(2), 139–151.
Cloke, P. and Jones, O. 2002. *Tree Cultures: The Place of Trees and Trees in Their Place*. Oxford: Berg.
Cohen, A. 1987. *Whalsay: Symbol, Segment and Boundary in a Shetland Island Community*. Manchester: Manchester University Press.
Feld, S. 1990. *Sound and Sentiment: Birds, Weeping, Poetics, and Song in Kaluli Expression* (2nd ed.). Philadelphia, PA: University of Pennsylvania Press.
Feld, S. 1996. Waterfalls of Song: An Acoustemology of Place Resounding in Bosavi, Papua New Guinea. In *Senses of Place*, edited by S. Feld and K. Basso. Santa Fé: School of American Research Press.
Feld, S. 2000. A Rainforest Acoustemology. In *The Auditory Culture Reader*, edited by M. Bull and L. Black. Oxford: Berg.
Franklin, A. 2006. *Animal Nation: The True Story of Animals and Australia*. Sydney: University of New South Wales Press.
Hartshorne, C. 1973. *Born to Sing: An Interpretation and World Survey of Bird Song*. Bloomington, IN: Indiana University Press.
Ingold, T. 2000. *The Perception of the Environment: Essays on Livelihood, Dwelling and Skill*. London, Routledge.
Ingold, T. 2011. *Being Alive: Essays on Movement, Knowledge and Description*. Abingdon: Routledge.
Ingold, T. 2015. *The Life of Lines*. Abingdon: Routledge.
Lorimer, J. 2007. Nonhuman Charisma. *Environment and Planning D: Society and Space* 25, 911–932.
Massey, D. 2006. Landscape as a Provocation: Reflections on Moving Mountains. *Journal of Material Culture* 11(1/2), 33–48.
Rapport, N. and Overing, J. 2000. *Social and Cultural Anthropology: The Key Concepts*. London: Routledge.
Whitehouse, A. 2017. Loudly Sing Cuckoo: More-Than-Human Seasonalities in Britain. *Sociological Review Monographs* 65(1), 171–187.
Wikan, U. 1992. Beyond the Words: The Power of Resonance. *American Ethnologist* 19(3), 460–482.
Wilson, K.-J. 2004. *Flight of the Huia: Ecology and Conservation of New Zealand's Frogs, Reptiles, Birds and Mammals*. Christchurch: Canterbury University Press.
Young, D. 2004. *Our Islands, Our Selves: A History of Conservation in New Zealand*. Dunedin: University of Otago Press.

6 "A feeling for birds"

Tuning into more-than-human atmospheres

Sara Asu Schroer

Escaping the churning heat of the midday sun we have sought refuge in the shade of one of the only trees around.[1] On fieldwork in Italy, I am helping Alistair with the training of two falcons, both youngsters that have not yet learned to fly. Raja, a peregrine falcon, is perched on my gloved fist. I can feel her strong grip through the leather as I observe her scrutinising the hillsides in the far distance, spotting things beyond my own awareness.[2] I try to adjust my body to her movements, broadening my perception to grasp that of the winged and feathered creature next to me. A gentle breeze of warm air is blowing up the hill. The bird immediately opens her wings, being gently lifted upwards by the rising air, before being lowered back down as the wind recedes. I adjust my fist with the direction of the wind to make her comfortable, trying to avoid the breeze blowing up her tail, which will easily throw her out of balance. The bird in turn ruffles her feathers and calmly stands up on one leg, ready for a nap. Outwardly remaining calm and steady, I am cheerful about my accomplishment. Even Alistair gives me an appreciative nod. "She usually is quite edgy when around people she doesn't know well, but she is comfortable around you".

Throughout the days when we were training the falcons Alistair instructed me that it was crucial to have a "feeling for birds of prey", referring to a certain sensibility towards the ways in which they experienced the world. Without such a sensibility the formation of a communicative bond with a creature that sensed and perceived the world in strikingly different ways to the human handler was said to be fraught with difficulty. Communication, in this context, is understood as referring to more than the transmission of information through certain media. Going back to its original meaning deriving from Latin *communicare*, to share, communication here points to the creation of affinity and a certain sense of community between humans and birds. Following falconers' narratives, I argue in this chapter, a central aspect of creating this sense of affinity is the ability to "tune in" to the particular atmospheric milieux in which the encounters take place. Developing an understanding of atmosphere as movement rather than constellation, it is here neither understood to belong to the realm of the purely emotional and psychic nor to that of the meteorological and physical but rather appears as a total spatio-temporal phenomenon that affects the ways sentient beings relate to others and their environments. This will be shown through the notion of "mood" that falconers employed. "Moods" were not described as emotions in the sense of inner subjective states, detached from the material world. The term referred, rather more holistically, to how a human or bird felt in, or was affected by, a specific situation, at once emotionally and physically. When interacting with a bird, "moods" could be sensed beyond the subject who experienced them as palpable atmospheric presence, sometimes described as "vibes" or "energy". The ability to attune to these moods was seen as crucial in order to build a communicative relationship with a bird and for the establishment of hunting cooperation later on. In the ending discussion, I will argue that the notion of atmosphere so far has been unproductively centred on human experience alone, suggesting to open it up to more-than-human perspectives. Here, I take inspiration from biologist and pioneer of ethology, Jakob von Uexküll (1934/2010), who highlighted early on the basic concept of mood (or Stimmung) as key for grasping how animals make sense of their subjective worlds.

I will begin with a brief contextual introduction to what falconers call a "feeling for birds", a notion that highlights the centrality in falconry of being open to the perceptive and sensorial lifeworld of another nonhuman being. I will then present two descriptive stories from my fieldwork, the first from the training of young birds, the second from the initial stages of creating a bond with a newly acquired bird. Both these stories show, in different but related ways, the significance that falconers attach to atmospheric phenomena such as the weather and moods, for successful interaction with their winged companions. I will, focus on how falconers understand the "mood" of their birds. Moods, like the weather, it will be argued, form part of atmospheric milieux that can be sensed beyond the subjective experience of an individual, human or nonhuman. The notion of atmosphere will here be used to illuminate the ethnographic material, seeking to get close to how the falconers experience and interpret their engagement with their birds.[3]

A feeling for birds

Falconry is a hunting practice in which humans and birds of prey learn to hunt in cooperation with each other. In the public imagination in Europe, falconry is often associated with the Middle Ages as a sport of the nobility. During my research many friends and colleagues were surprised to hear that falconry is still practiced today and that it has become a pastime pursued by people from different social backgrounds. In order to practice falconry in the UK, practitioners are not required to attain a licence as in many other countries. Usually novices are mentored by a more experienced falconer and learn through a form of informal apprenticeship. In falconers' narratives the birds themselves are described as active participants within these learning and teaching relationships, rather then merely beings acted upon by human handlers. When looking at these learning relationships, an aspect that particularly interested me was how two beings with very different perceptual abilities were able to reach a level of communication and understanding that made possible the development of a cooperative hunting practice (Schroer 2015; for the role of learning in human-animal relationships see also Anderson et al. 2017).

One of the elements that practitioners highlighted as important was to have a "feeling for birds of prey" – an ability that some described as impossible to acquire solely through learning and enskilment. It was seen by many as a more obscure aspect of the practice that was difficult to articulate. Having a "feeling for birds", in short, referred to an intuitive grasp of birds, to being open towards and somehow connected to the experiential world of raptors. This feeling, as falconers often pointed out, is an important precondition for establishing a bond with a bird of prey. It is, as one falconer stressed, about being able to "tune into the birds' world" and being receptive to how a bird might perceive the world in a specific moment. When asked about their experiences of discovering this affinity with raptors, a recurring narrative was that falconers discovered this "urge" or "passion" for falcons, hawks or eagles through a key revelatory event. Falconer Katy, from Wales, for instance, described the first time she had held a falcon on her gloved fist "as a moment when something clicked". At that moment she realised that this was what she wanted to do: "Working with a falcon, entering into her world and being able to make her feel comfortable around you . . . to make her trust you".[4] This openness, or feeling for birds of prey, was said to be necessary to create a positive bond with a raptor and to establish a shared sense of communication and understanding crucial for hunting cooperation. This does not however mean that falconry "is just something that comes naturally to you". As John insisted, "it takes a lot of learning and commitment to do the birds justice".

"Having a feeling for birds", then, was understood as a basis from which to build one's abilities as a falconer. During my fieldwork, falconers who were said to have this intuitive grasp of birds were very often described as quite withdrawn or 'awkward' in relationships with other humans. Indeed it is commonplace in falconry literature to describe people hunting with goshawks, in particular, as socially marginal characters who tend to take on the characteristics of their feathered companions (Cummins 2001: 220–221; Horobin 2004: 21). Though not always expressed in such vivid terms, falconers have nevertheless described their relationships with birds as transformative, shaping the ways in which they move their bodies and perceive the world. As I will argue in this chapter, a central aspect of this feeling was the ability to attune to the shifting atmospheric milieux, constituted by the forces of the weather as much as the moods of humans and bird.

"Soaring on rising air": tuning into the bird-in-her-environment

During my fieldwork in Italy we spent many summer afternoons training a young falcon for the upcoming hunting season. Raja was still learning to fly, and Alistair, the falconer I was working with, pointed out that it was important that she got to practice her flying skills in as many different wind and weather conditions as possible. At this stage of training Raja had started flying freely without any physical attachment to the falconer. Once the 'creance' was removed, allowing the bird to fly freely, human and bird were only kept close through a communicative bond that had been established progressively during the previous weeks (we will return below to an example of the early formation of such a bond).[5]

When training a bird of prey the choice of a suitable training ground is important and depends on the particular air currents in an area. These are dependent on the influence of the weather as much as on the lay of the land. For the early stages of flying Raja freely, Alistair chose a gently sloping hillside into which the wind blew moderately in the afternoon. The incoming wind meant that Raja would not be tempted to use a downward wind to fly to explore the valley (as might happen on the other side of the mountain). Alistair also warned that flying Raja during the hot midday was too risky, as there were thermal currents rising vertically from the ground that could tempt the young bird to soar up into the sky and drift too far off for the earthbound falconer to follow. Especially young birds were said to fall readily into a kind of trance, enjoying the lifting currents and forgetting about the pathetically waving falconer on the ground.

This initial stage of free flying is often described as one of immense anxiety and worry on the part of falconers. One of the main mistakes of falconry novices is said to be to prolong unduly the stage of training with the creance attached, out of fear of losing the bird. Alistair, reflecting on this typical experience, told me:

> It is not easy to make this step. It is quite contradictory in a way as you love this bird and spend all your time with her for weeks and yet then comes the day when you have to just let her go and hope that she likes you enough to stay close . . . at the end of the day there is always the potential risk to see the bird disappear into the distance, swept away on a downwind, never to be seen again.[6]

To avoid this, Alistair explained, one needs to be attentive to the way she feels in a particular situation, whilst at the same time trying to avoid one's own nervousness affecting the bird:

> When you are out with a falcon you need to really *tune in*, even though she might be flying high above you in the sky you should stay in close rapport, it is all about paying very close attention to the way she is experiencing the world up there. Hot days may make her moody; dry and cold air usually gives her a bit of an edge.

This example shows that 'having a feeling for birds' encompasses awareness of the bird-in-her-environment and with it the shifting atmospheric milieux in which we were immersed. A central dimension of such milieux is the weather and its various articulations that have direct influence on the moods or, as Alistair put it, "ways of experiencing" of the beings involved. Working with a winged and feathered creature brought the palpable yet transient forces of the weather within which we were interacting more clearly into my own awareness, forcing me to attend to how they influenced the ways birds (and falconers) responded to and experienced particular situations. Particularly for a bird in flight, variations of wind and weather had real consequences, whether in supporting flight or making it difficult, even dangerous, for her to manoeuvre. For us on the other hand, as earthbound humans who had to navigate on the rocky ground, it was a challenge to follow the aerial paths of the falcons above us (see also Senior 2016).

This point is emphasised by Alistair's use of the metaphor of "tuning in" when talking about his experience of handling birds of prey. When I asked Alistair to specify further what he meant by tuning in, he continued:

> Well, you need to get a sense for how your bird is feeling up there. It is not enough only to focus on the bird; you need to pay attention to what is happening around us and how it might immediately influence your bird . . . the weather is crucial . . . for instance humidity might slow her down, cold and dry air usually gives a bit of an edge. . . . It also has something to do with sensing the mood of your bird. One should really only start flying her freely when you are sure that the bird is in the right state of mind, I mean if she is particularly moody or edgy that day it might not be a good idea to try flying her as she might not respond well to you.

Through considering these examples from the initial training of a young falcon, we see that successful interaction between falconer and bird encompasses a heightened awareness of the atmospheric qualities of the environment. These encompass the forces of the weather as well as the affective, emotional presence of the sentient beings involved. In the following section I will develop this point through an example of the initial creation of a communicative bond between human and bird, whilst focusing specifically on the atmospheric phenomenon of moods.

"A bundle of nerves": manning a goshawk

I am watching Jonathan's female goshawk from a distance in the back of his garden. Recently arrived from a breeder in Germany, the hawk is getting used to her new surroundings and is perched, unhooded for the first time, on a block perch in front of a fence that shelters the garden from the park behind it. Without the hood, and in the bright sunlight of the day, her vision is clear and acute, and her responses are fast and immediate. Her body appears tensed; she is clearly not yet relaxed, her feathers pressing against her body, both feet clutching the covering of the perch. At times her attention is drawn to the agile swallows chasing insects above, at times to the restless movement of a dog exploring the bushes in the park behind her. Sensorially immersed, she is affected by the new sounds, vistas and movements around her.

The scene I am describing is from the so-called 'manning process', which in falconry refers to the initial stage in the developing relationship between falconer and bird. Jonathan, the falconer, had hesitantly invited me to join him to see his new arrival. Having had very little direct contact with humans in the breeding station, the bird was not accustomed to human company and everyday surroundings. These first weeks, in which falconer and bird were getting accustomed to each other, were very often out of bounds to me during fieldwork. This tends to be a very personal and private time in which falconer and bird are isolated from others while they gradually spend more and more time together and, eventually, with other people. Manning was described as a process of familiarisation, in which both falconer and bird learned to establish a basis for mutual understanding and communication. For falconers it was an emotionally challenging time, accompanied by fear and the worry of failing to establish a bond with the bird. In the falconry community, goshawks are often referred to as particularly sensitive and – compared to humans – 'nervous' creatures, with strong moods. They are described as "prone to panic attacks" and "shifts in personality", which can make them challenging to handle at times. Jonathan explained: "There is a certain intensity surrounding a bird that does not want to be touched or approached. A bird generally stays calm as long as you do not enter within a certain spatial distance from her"; it is crucial for the falconer to negotiate this personal space in the early process of developing a bond. As Jonathan continued: "It is almost as if when you do enter into that space you are not only touching her territory or personal space; you are also touching her".

As in the example above, this sensitivity to the bird-in-her-environment is central for falconers aiming to establish an initial bond with a novice bird in order to anticipate her responses and to avoid negative experiences. A central aspect of this is to sense the mood of the bird that is part in creating a certain atmosphere of encounter. As Jonathan explained to me before we went out to the garden:

> When a bird is sitting on a perch you need to be able to sense in what kind of mood she is, whether she is relaxed or intensely focussed and ready to bate at any moment. . . . It is a bit tricky to explain if you have not had the experience yourself . . . I guess it's a bit like a field of energy that surrounds a bird of prey.[7] When you are entering the space in which she is perched you have to be sensitive to the kind of vibes she is sending you. This field of energy . . . to go with that image . . . is flowing from you to the bird and back, and the way the bird responds depends on what kind of vibes you are sending out to her.

In a similar fashion to Alistair, who highlighted the importance of "tuning in" to the bird during training, Jonathan is talking about the need to be sensitive and responsive to the "vibes" and "mood" of a bird. This includes the need on the part of the falconer to closely control his or her own emotions and actions since birds are seen as:

> much more sensitive to what is going on in their environments . . . they will sense whether you are scared or relaxed and usually respond to things so much quicker than you. Sometimes I think they have the sixth sense . . . they just perceive so much more than I ever will.

Jonathan's explanation shows that falconers experience moods not so much as inner states, bound to the subjective experience of an individual, whether human or avian. Rather, I would suggest, moods were experienced as spatially tangible and palpable forces that influence human-bird interaction. Falconers often described this atmospheric character of moods through terminology such as picking up on the "vibes", "energy" or "intensity" of their encounters with birds. As described above the birds, in turn, were said to easily pick up on a falconer's nervousness, anxiety or anger, which had immediate consequences for the ways they would feel and respond. Part of the skill of falconry is therefore to if necessary be able to conceal one's true mood through controlling bodily movements accordingly. Moods, then, like the weather, are key to creating particular atmospheric milieux in which human and bird encounter each other. Let me return to my ethnography to elaborate on this.

From where I was sitting I saw that Jonathan decided to give it a go and to start approaching the perching goshawk, hoping to be able to make her step up on his gloved fist to feed. What sounds like a straightforward task is actually a delicate and difficult aim to achieve as it requires that the bird begins to trust the falconer and to feel comfortable enough in his presence to allow such close bodily proximity. Before we went to the garden Jonathan said he was quite nervous about this, as he had only once trained a goshawk before and that was many years ago. Yet, as he pointed out, it was important that the hawk should not sense that he was nervous, as this would immediately make her unruly too.

When commencing his approach towards the bird at the back of the garden Jonathan was superficially putting on an air of being rather unconcerned, while always remaining acutely aware of the bird and her environment. He did not approach her directly but was watering the plants in the flowerbeds on her right, an activity that she observed from a safe distance. He did so with calm movements, avoiding any rapid or loud manoeuvres. Progressively he was working his way forward until he was quite close to the perching bird, who, whilst still not completely relaxed, seemed to tolerate him in her close surroundings.

Eventually he slowly crouched down on his knees close to the perch, with a slightly forward-bent back he avoided towering high above her. With wings half spread out in a threatening gesture the goshawk started to hiss at him in furious defiance. Jonathan remained calm, crouched down a bit further and ceased approaching the bird. For a while he stayed in one place, his gaze averted and avoiding any rapid movements that might startle the hawk even more. He remained in this position until the bird ceased her threatening gesture before approaching further. With his right hand he reached out behind him, taking a piece of meat from a pouch behind his back, all the while attending to the bird in front of him, to whom he talked to in a calming voice. He then put the meat into his gloved fist and offered it to the bird, who, seemingly not yet sure what to make of this, looked at it, then at Jonathan, and back again. Jonathan lowered his arm a bit and began to talk to her, almost whispering. He encouraged her to take a bite but she refused and jumped off the bow-perch trying to bate away from him, wings beating fast and legs pulling on the jesses that were tethered to the block.[8] Jonathan outwardly remained calm – yet, as he told me later, he felt like screaming. He waited until the bird had calmed down and resumed her stand on the perch, leaving her alone for the time being.

Returning to the house he was a bit disgruntled, not so much with the goshawk as with himself.

> I should not have tried to feed her, she was clearly not relaxed enough and I should have seen it coming that she would try to bate away from me. Now it is best to leave her in peace a bit and to wait until her mood changes. Goshawks are funny that way: sometimes they are all fury and craziness and the next minute they are cute and playful like a little pet.[9]

Observing this scene it became evident that Jonathan and his goshawk were still in the process of developing affinity that would allow them to establish a sense of communication later on. A central aspect of this, was as has been shown above, the skilled attention to the dynamic atmospheric milieux through which human and bird engaged, a point to which I now turn in the ending discussion.

More-than-human atmospheres

Resonating with the observations made above, Philosopher Vinciane Despret (2004) has highlighted the centrality of attunement for the ways through which humans and other living beings become knowledgeable of each other's specific modes of being-in-the-world. For her this attunement involves a learning process to affect and be affected by other beings and worlds. According to Despret attunement enables "a new articulation of 'with-ness', an undetermined articulation of 'being with'" (Despret 2004: 131). This "being with" is also created by the larger situation of the encounter that helps to disclose or open up new possibilities of human-animal relating (see also Haraway on agility training 2008: 206–208). With regard to the fieldwork material, below I will argue that this larger situation can fruitfully be understood in terms of dynamic atmospheric milieux or a "force-field" (Edensor and Sumartojo 2015: 253) in and through which humans and birds learn to establish a communicative bond. I will subsequently show, following Uexküll, that mood – a subjective mode of human and nonhuman being – is a central dimension of such milieux.

Atmospheres have often been described as belonging to the realm of the 'in-between', troubling analytic distinctions that work with dualist categories. In a recent edited collection, for instance, Bille et al. suggest that when we experience an atmosphere, we may not be sure "whether we should attribute them to the objects or environments from which they proceed or to the subjects who experience them" (2015: 2). Atmosphere, in this understanding, is neither a quality of the world 'out there' nor a psychic state internal to the individual.

This view of atmosphere, that is dominating much of the current debate on the social and material production of atmosphere, is most prominently associated with the work of phenomenologist Gernot Böhme. His approach to aesthetics locates atmosphere in the interstitial spaces in-between things, or between object and subject. For Böhme atmospheres arise from the sum total of what he calls the 'ecstasies of things'. By its 'ecstasies' he means the ways in which a thing 'goes forth' from itself, that is, how it is sensuously present in a certain situation (Böhme 1995: 32–33; Böhme 1993: 121–122). Atmospheres, then, are the "perceived quality of a situation", constituted by a specific "constellation of things and people" (Bille et al. 2015: 32).

With regard to understanding the atmospheric milieux of falconry practice, this understanding of atmosphere is both helpful and limiting. On the one hand, examples such as Jonathan's training of the goshawk show that atmospheres are indeed experienced as having a spatio-temporal presence that exceeds the subject experiencing them, whilst also not being independent of the sentient beings involved. My material does therefore suggest an understanding of the atmospheric that is not easily accommodated within frameworks that presuppose an a priori boundary between humans and their 'external' environment, but rather demands a more fluent and permeable conception of human-environment relations. However, Böhme's constellationist approach becomes less convincing in its identification of things and people as the *source* of atmospheres, that are subsequently to be located in-between the perceiver and the perceived (see also Riedel 2015 and this volume). This relatively static view does not leave room for what falconers highlighted frequently, that is the dynamic and transformative effects of atmospheric phenomena – such as moods and the weather – upon how humans and birds relate to each other and their environments. In contrast to Böhme who understands atmospheres as properties of the in-between, I would therefore, rather suggest thinking of them as spatio-temporal milieux *within* and *through which* humans and birds become attuned to each other.

Such a movement-focused understanding of atmosphere finds support in the work of phenomenologist Schmitz. He criticises Böhme for suggesting that atmospheric movement occurs only in between already established subject and object positions, suggesting instead that subjecthood or objecthood be best understood as phenomena that do not precede but rather emerge from atmospheric currents (Schmitz 1998; Riedel 2015. Similarly, Ingold highlights the generative character of the atmosphere, which he sees not as set over against the perceiver as "targets of perception" but rather as a transformative medium through which people perceive (Ingold 2011: 134). He writes: "Light, sound and feeling tear at our moorings, just like the wind tears at the limbs of trees rooted to the earth. Far from being enfolded into the body . . . they take possession of it, sweeping the body up into their own currents" (Ingold 2011: 134–135). By giving primacy to movement, Ingold asserts that "the wind is its blowing", not an object that blows, as much as a person is what she is doing rather than a being endowed with agency. In this sense, atmospheres are not the outcomes of interaction between already existing entities but are rather movements in themselves, "always in the process of emerging and transforming" (Anderson 2009: 79).

As we have seen in the examples of training and manning birds of prey, both the weather as well as the moods of human and birds play part in constituting this dynamic atmospheric milieux in which they interact. Moods, in particular, constitute a dimension of atmospheres that is of paramount importance for the establishment of a communicative relationship between falconers and their birds. More-than-human aspects of atmospheres, however, are often neglected in the literature, possibly due to its largely phenomenological borrowings. The work of Jakob von Uexküll is an exception to this as he attempts to develop an approach to meaning production beyond a purely human-centric perspective (Uexküll 2010). Interested in exploring nonhuman lifeworlds, Uexküll uses mood as a basic concept to understand the perception and behaviour of nonhuman beings. Notwithstanding the complexity of a life form, he argues, moods are crucial for the way in which an animal interprets its environment (ibid. 92–95 and Portman 1956). Moods, for Uexküll, are subjective and belong to the 'inner' lifeworld of living creatures. As such they cannot be observed directly but only inferred to through the observation of the animals activities. Von Uexküll depicted these lifeworlds, or *Umwelten,* as perceptual spheres, enveloping living beings according to their species-specific forms of perception (ibid.: 2–3). What I want to take from von Uexküll's work at this point is his claim to establish mood as a basic concept for human *and* nonhuman meaning-making. This understanding resonates well with the lived experience and narratives of falconers, described above, whilst also needing to be opened up to accommodate the atmospheric and spatially tangible dimension of moods.

In anthropology, moods have not so far been a topic of extensive scrutiny. An exception is the work of Jason Throop (2014), who identifies moods as atmospheric phenomena that are "intermediary forms of experience", lying between the "polarities of conscious ontological reflection and embodied disposition" (ibid.: 70). Following Valentine Daniel he understands moods as a "disposition toward the world" from within which humans reflect on life and moral concerns (Daniel 2000: 333 quoted in ibid.: 70). With this understanding Throop builds on the phenomenological writing of Heidegger, for whom mood (in German, *Stimmung*) constitutes a foundational aspect of human being-in-the-world.[10] According to Martin Heidegger people are always in one mood or another, and according to that mood the world will open up in a particular way. Moods, however, are understood as bound neither to the subject nor to an external world but as arising "from being-in-the-world itself, as a mode of that being" (Being and Time 1996: 129). It is this world-disclosing characteristic of moods, I would argue, that also becomes important in the social interaction between falconers and their birds. For them, in contrast to the anthropocentric stance of much writing in phenomenology, moods are not bound to human experience alone but are also an aspect of nonhuman existence. As Alistair explained: the "tuning in" to the bird's mood, or the "feeling of vibes" when approaching a perched bird, is a central requirement of being able to successfully communicate with a creature that senses and perceives the world in different ways from those of the human handler. Communication, I would suggest, reveals itself not as a straightforward exchange of signs easily acquired by anyone – given the right kind of instruction and time to learn – but also hinges on an attunement to the mood of the other: to the atmospheric milieux of the encounter.

Conclusion

Based on fieldwork with humans and birds of prey in falconry practice this chapter aims at opening the discussion on atmosphere to more-than-human perspectives. It has been shown that phenomena of atmosphere are central to the ways in which humans and birds learn to communicate and experience each other and their environments. Building on the notion of attunement put forward by Vinciane Despret, I argued that the atmospheric milieux in and through which living beings encounter each other is crucial in providing the conditions for inter-species affinity and communication to emerge. Arguing for a movement-based understanding of atmosphere, I have shown that part of the dynamics of atmospheric milieux are moods, spatially tangible beyond the subjective experience of the individual. In falconry practice the 'fury', 'calmness' or 'nervousness' of a bird was not bound to the idea of an inner psychic state but could be felt atmospherically in bodily encounters as 'tension' or 'energy'. Falconer Alistair, for instance, described his experience of the centrality of moods as one of 'getting the vibes' of his falcon flying high above him in the sky. This 'tuning in' was their basis for creating a communicative relationship in which they both became participants in a shared social world. Meaning, interpretation and hence communication, then, are only possible because living beings are essentially moody creatures; always already attuned to the world in a situation-specific way.

Notes

1 The work on this chapter has been made possible through a research fellowship as part of the ERC funded project Arctic Domus at the Department of Anthropology, University of Aberdeen. I also would like to thank Louise Senior, Tim Ingold and Hakon Caspersen for their comments on drafts of this chapter.
2 Falconers differentiate female and male birds by personal pronouns. By default, birds are designated as female if the sex is not known or if the reference is to the birds in general. In this chapter I will adopt this usage, rather than referring to the birds as 'it'.
3 This chapter is draws on my Ph.D. research into the practice of falconry. This research was based mainly on ethnographic fieldwork in Britain as well as smaller fieldtrips to Germany and Italy (co-funded through the International Rotary Foundation, Sutasoma/Radcliffe Brown Trust of the Royal Anthropological Institute, the Principal's Excellence Fund University of Aberdeen, Falconry Heritage Trust and Deutscher Falkenorden).
4 Similar narratives can be found in memoirs and personal reflections of falconers, see for instance Glasier 1963 and Gallagher 2008.
5 Creance is a falconry term for a long leash used in training of birds. The creance is tied to leather anklets and swivels at the bird's ankle as well as to the falconer's leather glove. In training, when the bird learns to fly to the falconer for food, the distances between falconer and bird, and the length of the creance, are gradually increased until the bond between human and bird is developed enough to allow free flying.
6 Nowadays it is common practice to fly birds with the use of telemetry receivers that allow the tracking of birds over long distances. This reduces the risk of losing a bird significantly and is used as an addition to falconry bells that have been traditionally utilised to locate birds that have flown out of sight.
7 To bate refers to when the bird beats the air with her wings, often used to describe a bird flying off the fist or perch, or attempting to (when tied), either in the pursuit of

prey or because she has been frightened, baiting can therefore be seen also as a sign of displeasure.

8 Jesses are thin strips of leather attached to anklets on the bird's legs used to hold the bird on the gloved fist by the falconer.

9 For a recent publication that echoes some of the observations made on the particular relationships between falconers and goshawks see Helen Macdonald's autobiographical novel *H is for Hawk.*

10 For literature dealing with the influence of von Uexküll's work on the development of phenomenology, see Buchanan 2008.

Bibliography

Anderson, B. 2009. Affective Atmospheres. *Emotion, Space and Society* 2, 77–81.

Anderson, D., Loovers, J. P. L., Schroer, S. A. and Wishart, R. P. 2017. Architectures of Domestication: On Emplacing Human-Animal Relationships in the North. *Journal of the Royal Anthropological Institute* 23, 398–418.

Bille, M., Bjerregaard, P. and Sørensen, T. F. 2015. Staging Atmospheres: Materiality, Culture, and the Texture of the in-Between. *Emotion, Space and Society* 15, 31–38.

Böhme, G. 1993. Atmosphere as the Fundamental Concept of a New Aesthetics. *Thesis Eleven* 36, 113–126.

Böhme, G. 1995. *Atmosphäre*. Frankfurt am Main: Suhrkamp Verlag.

Buchanan, B. 2008. *Onto-Ethologies: The Animal Environments of Uexküll, Heidegger, Merleau-Ponty and Deleuze*. Albany: SUNNY Press.

Cummins, J. 2001. *The Hound and the Hawk: The Art of Medieval Hunting*. London: Phoenix Press.

Despret, V. 2004. The Body We Care For: Figures of Anthropo-Zoo-Genesis. *Body and Society* 10(2/3), 113–134.

Edensor, T. and Sumartojo, S. 2015. Designing Atmospheres: Introduction to Special Issue. *Visual Communication* 14(2), 251–266.

Gallagher, T. 2008. *Falcon Fever: A Falconer in the Twenty-First Century*. Boston, MA: Mariner Books.

Glasier, P. 1963. *As the Falcon Her Bells*. London: Heinemann.

Haraway, D. J. 2008. *When Species Meet*. Minneapolis, MN: University of Minnesota Press.

Heidegger, M. [1927] 1996. *Being and Time*. Albany: State University of New York Press.

Horobin, D. 2004. *Falconry in Literature: The Symbolisms of Falconry in English Literature from Chaucer to Marvell*. Surrey: Hancock House Publishers.

Ingold, T. 2011. *Being Alive: Essays on Movement, Knowledge and Description*. Routledge: London.

Macdonald, H. 2014. *H Is for Hawk*. London: Cape.

Portmann, A. 1956. Ein Wegbereiter der neuen Biologie. In *Streifzuege durch die Umwelt von Tieren und Menschen: Ein Bilderbuch unsichtbarer Welten*, edited by J. von Uexkuell and G. Kriszat. Bedeutungslehre. Hamburg: Rowohlt, 7–17.

Riedel, F. 2015. Music as Atmosphere: Lines of Becoming in Congregational Worship. *Lebenswelt* 6, 80–111.

Schmitz, H. 1998. Situationen und Atmosphären: Zur Ästhetik und Ontologie bei Gernot Böhme. In *Naturerkenntnis und Natursein: Für Gernot Böhme*, edited by G. Böhme, M. Hauskeller, C. Rehmann-Sutter and G. Schiemann. Frankfurt am Main: Suhrkamp, 176–190.

Schroer, S. A. 2015. *'On the Wing': Exploring Human-Bird Relationships in Falconry Practice*. Ph.D. thesis, University of Aberdeen.

Senior, L. 2016. Energy Use as Skilled Process: Gardeners and Wind in Highland Scotland. *Anthropology Today* 32(4), 3–7.

Throop, J. C. 2014. Moral Moods. *ETHOS* 42(1), 65–83.

von Uexküll, J. [1934] 2010. *A Foray Into the Worlds of Animals and Humans: With a Theory of Meaning*. Translated by. J. D. O'Neil. Minneapolis, MN: University of Minnesota Press.

7 Making charismatic ecologies

Aquarium atmospheres

Susanne B. Schmitt

An aquarium shop in a medium-sized German city. Just meters away from a noisy, dusty, multi-lane street, you enter a world of calm, dimmed light and delight. This is not the most glamourous part of the town, Munich. Outside, lunchtime draws the working crowds to a quick fix of Kebap and Falafel. Post-war concrete architecture holds little pawn-shops, small boutiques with Chinese textiles, and secondhand clothing stores. Carpet stores with '50 % Sale' signs that have been up for ages, bleached by many summers. It is hot and dry outside; the sunlight mercilessly shows every oil spot on the asphalt, every spit out and hard-packed chewing gum and cigarette butt. So bright is the sunshine that the large shop window, specially coated against sun rays, appears almost opaque. Behind the shop window though, if you move really close, is the Amazon river.

Hundreds of neon tetra *[Paracheirodon innesi]*, glittering clouds, move quickly amidst a lush underwater forest, changing direction at tiny signals that are incomprehensible to a human non-swarmer. Amidst them, floating gently, pairs of angelfish *[Pterophyllum scalare]* slowly sail across their territory. There is some serious roadwork going on outside. A group of construction workers have torn open the heated asphalt; their shapes, their bright orange protection gear are mirrored in the tank walls. The harsh, unbearably loud noise of the jackhammer sends a dystonic rhythm out onto the street. The oily dust they stir up settles on the sweaty summer skin of the passerby in a heartbeat.

On the tank walls inside the shop, the orange safety gear, my own shape in front of the shop window, and the lush underwater jungle in shades of green and rust mingle like the staggered scenes on an old-fashioned theatre stage. It is cool inside; the air is pleasantly fresh, enriched by the fresh and salty waters from more than forty fish tanks. These tanks present a diverse spectrum of underwater sceneries to the onlooker and potential customers. The shop, a labyrinth of metal shelves with packaged filters, heaters, sand, pebbles, and aquarium plant fertilizer, is shot through with islands of subaquatic light. Not all of them are as big as the shop window basin that holds the Amazon riverbed scenery. Many are no larger than a package of orange juice and reveal the life within, tiny and frail invertebrates, snails and freshwater snails, only to a patient and persistent gaze.

There are the large, limestone-covered tanks with hard water where territory-building, large rock dwelling blue and yellow cichlid, predatory fish of the East

African lakes, thrive. Their underwater jungle is a petrified one: large chalkstones with no plants except for the periodical head of lettuce they have for dinner. When they breed (and here, they do, because they are provided good conditions), they keep their offspring in their mouths for shelter amongst their rivalling, hungry neighbours. South East Asian aquariums with different species of gouramis are dipped in a duskier light; the tanks shadowed by water lentils and lotus floating on the water's surface: slow swimmers who cling together in small groups, these so-called labyrinth fish are often orientated towards the water surface where they, fish for whom the oxygen dissolved in the water is not sufficient, can breathe air, hide, and lay their eggs in the roots of the floating water plants.

Introduction

I came here for some filming, and for some talking, too. I came here, more precisely, to learn about the making of aquarium atmospheres from an experienced aquarist whose family business has built and maintained home aquariums and larger public displays for two generations. In this chapter, I am looking at the making of aquarium atmospheres. I do so by interweaving what I learned from Max Waldhäusler, the shop owner, and other aquarists, my own 20 years of experience in the designing and making of aquatic environments, and historical trajectories of creating them that I take from literature. I especially look at practices of "orchestrating" aquarium atmospheres with light (Bille 2015: 259), and at processes of making "good water" – the aquarium's atmospheric element. By doing so it will be shown how creating aquariums that living organisms can inhabit and apprehend is a human project of the creation of atmospheres in a twin sense – an aesthetic and a physical one (McCormack 2008; Ingold 2012: 80; Ingold 2015; 73). I contribute to the question of the making of atmospheres (see for example Edensor and Sumatojo 2015; Pink and Mackley 2016) by stressing how a turn towards the "key roles of subjects in co-producing atmospheres" (Edensor and Sumatojo 2015: 2) needs to include more-than-human subjects, experiences and species dependency on atmospheric forces like air and water.

In fact, the challenge of making aquariums – building a durable environment for organisms whose needs and perceptions aren't your own – has been a crucial epistemic and practical stepstone for the development of ecological theory by the likes of Konrad Lorenz and Jakob von Uexkuell (Wessely 2013). Why detailing this strand of scientific history is beyond the scope of the chapter, it does take the notion of ecology seriously: ecological thinking is a cornerstone of atmospheric making in aquariums. I thus propose the notion of 'charismatic ecologies' for them, borrowing from Jamie Lorimers (2015) notion of nonhuman charisma, which entwines the corporeal properties of an lifeform, the possibilities for humans to experience them and enter into their world, and the affects engendered in those close-up, multisensory encounters. This kind of charisma is spread out in aquarium habitats, moving beyond single charismatic organisms.

Aquariums, tuned spaces and dwelling sites for other-than-human life have a history that entangles human aesthetic ideas and imaginaries with the manipulation of wayward and fragile elements – here, water. Aquariums are micro-worlds that entwine scientific fascination with aesthetic pleasure where the species and specimens they contain "became metonymic of particularly nonhuman environments" (Hayward 2012: 165) and brought with them a "promise of immediacy" (Hayward 2012: 167) of more-than-human encounters – from the 19th century until today (Olalquiaga 1998). They are enchanting sites for human onlookers whose very material conditions are crucial for the survival of the many precarious, interconnected species that inhabit them. Aquariums are thus more than representations – they are sites where forms of life "stitch themselves together" (Ingold 2015), weaving together strands of different material and aesthetic needs and potentials.

Entering aquarium atmospheres

If I talk about an aquarium atmosphere, you may immediately know what I mean. Within the "familiar taxonomy of atmospheres" (Anderson and Ash 2015: 36), atmospheres that radiate from aquariums are regularly addressed and represented in everyday language. Not all atmospheres can be easily named. What sounds at first unsurprising is in fact both peculiar and important. As Tonino Griffero points out, to mention the 'atmosphere' of a setting or of a situation usually implies something positive. To say that something has "an atmosphere", a restaurant for example, is most often a compliment (Griffero 2015). Identifying and explicating atmospheres serves the purpose of stabilising them and making them representable (Anderson and Ash 2015) – and in the case of public aquariums and the pet fish industry, evoking aquatic atmospheres is tied to a billion dollar business. In keeping with Ben Anderson and James Ash, however, I think of naming atmospheres "as part of a methodological practice involving a combination of description and speculation". Even if they may be slippery, there is something to be said about aquarium atmospheres, historically and contemporarily.

In colloquial, everyday conversation, the aquariumness of an atmosphere may implicate the aquarium's glass walls that separate the classical atmospheric medium, air, from water. In that case, an aquarium atmosphere might be the slightly uncanny one of being observed – like, say, in an airport's smokers' lounge, a glass cubicle slowly filling up with cigarette smoke that both hides and exposes those inside to outside gazes.

Little seems necessary to evoke an aquarium atmosphere – or to charge the atmosphere of an event with subaquatic quality. The light, greenish or bluish, matters a great deal, as we will see again later. Even the traces of the forms of life that inhabit those glass constructions suffice to gesture towards the aquarium as a complex micro-ecology and dwelling site. An example: for a moment in 2012, there was a stage set made by the recently deceased, world famous architect Zaha Hadid.[1] It was all kept in white. Its walls, where the audience was placed, shimmered in iridescent mother-of-pearl. Large mussels, seaweeds, a jellyfish, made of shiny white synthetics, surmounted the models moving between them like a school of small fish. This aquarium served as the runway of Karl Lagerfeld's Spring/Summer 2012 fashion show for Paris at the Grand Palais during Paris

Fashion Week. Singer Florence Welsh, standing on a shell, performed 'What the water gave me'. The shimmering of light, brought about by the pearly fluorescence, the shapes of seabed creatures, were enough to hint towards aquariumness. Some elements were sufficient to evoke a whole, they gesture towards it (see Strenslund's chapter in this volume on gesture).

While little is necessary to give a site a hint of aquariumness, the feel of being in an aquarium, turning glass cubes into micrological ecosystems where life can flourish is quite a different matter. In what follows, we will thus look at what making aquarium atmospheres entails from a more-than-human perspective.

Aquariums as charismatic ecologies

Most aquariums are made to look pleasing to humans, but although the practice has long connection to playful *kitsch* (Olalquiaga 1998), idealised naturalism has by now become its most important design trope. Many home aquarists combine fish, invertebrates and plants from different world regions. Their needs and environmental requirements, however, need to be compatible, and aquaristic guidebooks come with detailed description of the water quality, general layout, food requirements and ways of life of specific species to help aquarists determine if certain organisms can live together in a man-made tank. Whether tanks mirror situations that can be found in settings that are imagined as wild or constructed by help of a guidebook, balance is key.

While decorativeness, for the inhabitants of a tank, is not needed to stay alive, ecological possibility is. And yet aquariums need to have aesthetic appeal as well. I thus call them "charismatic ecologies", a notion that I borrow from Jamie Lorimer's notion of "ecological charisma" (Lorimer 2015). It moves beyond the well-known concept of 'charismatic species', which Lorimer takes from conservationists. He uses it to address species with much popular appeal (2015: 39) and a great deal of aesthetic and corporeal, usually visual, but also aural, charisma: megafauna like the Asian elephant are a good example, and their public appeal makes them a node for conservation efforts by proxy for other, less appealing species, who still profit from the protection of ecosystems that their more charismatic cohabitants dwell in. Not only species can be charismatic, but whole ecosystems can be charged with specific affective logics that present them as happy, full of good atmosphere and thus worthy of fascination and even protection. Aquariums here serve as "sensory prosthetics" that "expand the scope of the human *Umwelt* and help overcome many of the perceptual constraints" (Lorimer 2015: 44) that living through different elements usually presents to earth-bound and aquatic species. When I call aquariums charismatic ecologies, I mean precisely this: they give each species what it needs, and serve as a unifying envelope for shared apprehension.

The aquarium phenomenon, these charismatic ecologies, oscillate between home and large public versions, and both take inspiration from each other: aquariums in zoological gardens or commercial setups would be impossible without the experimentation of amateur scientists and home aquarists that were crucial for understanding how underwater lifeworlds best be made. Home aquariums, on the other hand, often take inspiration from the setups of large public aquariums and their strategies of display. In the next subsection, we will look at how these setups came into being.

Historicising aquarium atmospheres

A different scenery: In the woods north of London, a group of ocean activists – conservationists, journalists, aquarium professionals – meet for a workshop by the name of "Circumnavigating Hope".[2] This workshop is aimed at finding new ways of communicating 'ocean optimism', positive stories about conservation instead of end of world scenarios – an idea that will later be turned into a twitter hashtag with millions of followers.[1]

During the workshop, a curator of the Plymouth Aquarium shares his profound fascination for aquarium exhibits with me: in a zoo, working with birds or mammals, he explains, you can exhibit a single species and provide it with the best environment possible. In an aquarium, on the other hand, you are presenting a window "into a whole world" as he puts it: an ecosystem with all its interdependencies of life forms, multiple layers, and what he sees as creative possibilities. This is part of their success. They are more than just backgrounds for fish. Knowledge about the interdependencies at play in a tank are key to making and sustaining one.

Max at the aquarium store in Munich said something very similar when we discussed what was so special about making and owning[3] an aquarium:

> Everybody knows a dog, a cat, a horse . . . no matter what animal you keep, you always lock them in [. . .] With the aquarium the thing is, the fish gets in there and has no way out. This is why you should make sure the fish is doing as well as possible, as it would in nature. Changing conditions, good water, good environment, things like that. On the other hand I find it important to engage with nature . . . how the animals react, that this is an ecosystem, which is running on its own, that if there is too much external influence, things go into a spin.

From Shanghai to Cape Town, from Munich to San Francisco, aquariums, although locally adapted, put similar tropes forward and utilise similar techniques to avoid things going, as Max put it, into a spin.[4] This is because aquariums come with an aesthetic history of their own, a history that is tied to the notion of ecological relationality and connectivity my interlocutors talk about. Their atmospheres became appealing at a time when urban elites of the Global North discovered science as a pastime, and oceans became a well of fascination rather than fear (Brunner 2005).

In the Euro-American West, aquariums have existed since the late 19th century (Olalquiaga 1998). During the 1840s and 1850s, Great Britain experienced an aquarium craze that soon swept over to continental shores. Before the aquarium became a public spectacle and decorative centerpiece of bourgeois homes, most land-bound humans had very rarely even seen undersea species with their own eyes (Brunner 2003). The oceans were regarded as highly dangerous and as inhabited by monsters – an assumption that was all too often fueled by the giant cadavers of octopi, jellyfish or even whales that had been washed onto European shores (Brunner 2003: 9). In addition, it was commonly believed that the water pressure in the ocean was so high that no life could possibly exist on its deepest ground. Advances in diving technologies proved this assumption wrong, and at the same time, the development of tourism turned beaches, previous uncanny and uncomfortable places, into new sites of relaxation where the bourgeoisie was confronted with what was washed ashore – and even started to enjoy some leisurely strolls alongside tidal pools and its other-than-human inhabitants (ibid.). Naturalism – including the collection of botanic and zoological specimens – was hugely popular at that time, and seaweeds and sea grasses were the first aquatic specimens that ended up in the home, either in a glass jar, or pressed, in an album. Flourishing businesses began to provide customers with animals, plants, and fresh seawater. Before aquariums finally made their way into human habitats though, one had to find ways to stabilise them and make them sustainable environments where life was possible. Aquariums needed to have an atmosphere in the strictly climatic sense of the term. The right amounts of oxygen and carbon dioxide needed to be dissolved in the water, which was possible only if plants and animals coexisted in a perfectly balanced state of biochemical relativity: this was the meaning the word "atmosphere" carried in the aquarist contexts of the day – the aerial part of gas exchange. Only when it was understood that oxygen and carbon dioxide together created a liveable underwater climate could the aquarium craze really take off (Olalquiaga 1998: 51). The very concepts of 'ecology' and 'milieu' go back to this historical moment: around 1900, the emerging field of aquatic biology, featuring masterminds of environmental thinking such as Jakob von Uexküll, began to understand the complex relationships between individual organism and environment (Wessely 2013). Aquariums were ideal for experimentation with this new mode of thought and the very epistemic objects that helped to formulate ecological thinking about interdependencies and connectivity in a controllable settings (ibid). In a laudatio in 1980, looking back at his career, Konrad Lorenz, one of the founding fathers of ethology, claimed that the aquarium was one of the most important objects for the development of his thinking. In a statement not too different from what contemporary aquarists like Max would say, he noted that "there you have . . . an ecosystem in front of you eyes that is flourishing only as a whole, or the fish that you intend to study scientifically is gone" (in Wessely 2013: 1, translation by the author).

Especially to those who kept aquariums at home, and to the businessmen who were keen on opening public aquariums, it was not only the undersea ecologies and their physical necessities that mattered. Atmosphere in a decorative sense did, too, and they needed to be habitats engendered with charisma. Aquariums of the time were richly decorated vessels that sat enthroned on antimacassars. Mostly without covering, the plants they homed grew above the glass walls and were often complemented by terrestrial plants on the coffee table that continued the idyllic situation onto dry space. To the contemporaries, the aquarium spoke not only of a certain wealth but also of scientific appreciation cum upper class taste. The aquarium was not only an atmospheric bubble in itself, the recreation of an underwater scenario – that of the 'coffee table lake' or 'the sea in a glass jar' (eg. Harter 2015). Its boundaries were fluid. It affected the whole living room.

Creators of public aquariums tempted the boundaries of the glass cube even further. The Berlin aquarium, placed at the popular shopping and strolling promenade 'Unter den Linden', opened in 1869, and invited visitors into 'Neptune's realm' itself. Like other public aquariums of the time, the design trope used was the one of the grotto (Olalquiaga 1998). It created an all-encompassing sense of being on the ocean ground, the very place one had, in 1869, believed was completely lifeless a mere 40 years ago. At 'Unter den Linden', Visitors drifted along pathways through a dark network of caves and grottoes where they passed through various geographic zones inhabited by local sea animals, gazing into their lit-up tanks. Birdsong, the shrieks of apes and the sound of water accompanied them through this deliberately multi-sensorially designed place whose first executive director and initiator, Brehm, was a trained architect (Harter 2015: 42–57). A then new technique, indirect lighting, added to the sense of being in an enchanted underwater wonderland. The tanks were partially lit by sunlight guided along complex constructions as well as by electricity. Local newspapers advised Berliners to visit at least twice, by sunlight and when it was dark, so that they could capture the different moods and activities of nocturnal tank dwellers and those active during daylight hours (Harter 2015: 47). Human aimless wanderers added to the peculiar feeling of being in a submarine space. From their writings, we know that the presence of others while they gazed into the private life of marine animals created a special sense of intimacy and in-betweenness – a happening of swarming bodies, as the journalist Dorothee Goebeler recalled 110 years ago, in 1907, in a Berlin newspaper:

> I did not pay attention to my steps, in my ears I heard the quiet sound of the sea, I saw the rising tide and the foaming surf; right in front of me was an unfamiliar world in motion, it was swimming in front of my eyes like a kaleidoskope (sic!), the ocean shone for miles around with electrical light, fish criss-crossed the green sea, [. . .] and the brown anemone waved and brought me back to the submarine meadow where the anemones silently bloom.
>
> (In Brunner 2005: 111)

Making good water

At the aquarium store, I now sit on a leather bench in front of the Amazon river landscaped aquarium which is several meters long. With me is Max Waldhäusler. Max has learned the trade from his parents. The family business has been catering to Munich-based aquarists for two generations. The results of their work can be admired at the brand spaces of high-end cosmetics companies like 'Creme de la Mer', at bars and restaurants, and offices. Especially on Saturdays, the store that appears so unassuming on the outside is buzzing with amateur aquarists – clients who equip themselves for a weekend of aquaristic bliss, of setting up, redesigning and caring for their often numerous fish tanks over the weekend. They are here to change the mood, the *Stimmung* of their own human homes into places of calm and curiosity where yellow-green freshwater light or the bluish white of marine waters disgorges from a tank in the corner of a room.

Creating aquariums that living organisms can inhabit is a project of the creation of atmospheres in a twin sense – an aesthetic and a physical one (McCormack 2008; Ingold 2012: 80; Ingold 2015: 73). Air, or Greek *athmos*, the word we lend the notion of atmosphere from, is the classical medium of such inhabiting: "It is with our entire being, indissolubly body and soul, that we breathe", and the acts of breathing are how humans are tied to atmosphere (Ingold 2012: 83). Just like air, water is a total, all-encompassing element. In order to have fish and other aquatic species live at human homes, the chemical relations of the aquatic element need to be balanced correctly. Filtration, aeration, water changes and circulation are crucial for keeping the water a sustainable envelope for aquatic organic forms.

Plants, snails and other forms of life need to live in equilibrium amongst themselves in complex ecologies so that water can become an element that brings life, and not death. Accordingly, aquariums cater to different needs that must always come together in a symbiotic and sometimes confrontational synthesis of life worlds – or rather, of worlds where life is possible: they must be places for aquatic species to live and thrive in, and they need to cater to the decorative claims that aquarium owners lay on them. A shopping list for a Waldhäusler aquaristic store visit is thus often the result of tedious research on the living conditions and needs of the future inhabitants. Electrical heating, a filter, fertilizer, sand and pebbles, a thermometer, and of course a tank with lighting are on the beginner's' shopping list. A few plants, robust and fast growing ones, (assuming you are building a freshwater basin) to begin with. Then you can start building, layering fertilizer, sand or pebbles on the ground. Every material needs to be chosen with the eventual outcome in mind. You want an Amazon river basin, like the one in the shop window of the aquarium store? Then, you aim for very soft water, no limestone rocks, and many roots.

For an Amazon river Blackwater stream aquarium (which is what you want if you want what Max has), construction advice reads as follows[5]:

> ***Blackwater Stream*** = South American blackwater streams originate in the rainforest and thus have a great deal of accumulated leaf litter. Because the water is slow-moving, acids (called tannins) are continuously leeched from the decaying vegetation which gives the water a transparent tea-colored hue. The water in a blackwater stream is typically soft with an acidic pH between 4.5 and 6.5. The recommended substrate for a blackwater stream tank is fine clay or sand covered with a few inches of leaf litter. Lighting should be subdued and filtration somewhat slow.
>
> *Recommended Species*: Angelfish, Discus fish, Tetras, Corydoras catfish, Dwarf cichlids, Hatchetfish

Water, the aquarium's atmospheric element, takes many weeks to become "good water": a balanced medium where microbes and bacteria that travelled along with the roots, plants and leaf litter have out-balanced each other and created a sustainable element.

For aquarists of an earlier generation, rain was the way good water became available to them: soft, without hard-to-remove limestone, and fresh. They used to set out buckets when rain clouds came up. In times of increasingly polluted "changed air" (see Choy and Zee 2015), however, rain is no longer to be relied on. Nowadays, well-monitored municipal tap water becomes aquarium water. Filter sponges from well-established aquaria make great gifts at the aquarium store and among networks of aquarists: you can't buy them, but you may be given them. With them, Max knows as well as any other seasoned aquarist, 'good' bacteria travel to the newly set up tank. They process the organic waste that is soon collecting at the aquarium. They make 'good water', a good physical atmosphere. Through their activity, waste becomes Ammonia. Other bacteria turn Ammonia into Nitrite. Another set of 'good bacteria' transforms Nitrite into Nitrate. This process is called aquarium cycling or the aquarium nitrogen cycle.

A trained aquarist's eye knows pretty well what the water is like at any moment: its colour keeps changing ever so slightly, and 'good water' for a blackwater tank has a very specific, slightly yellow glow. A new, organic smell emerges. By now, small ramshorn snails [Planorbarius corneus] will have hatched. They travel along with the plants; there are always some. If they thrive, for now at least, all is well.

After 10 days, or maybe two weeks, an event visible even to the untrained eye takes place: the "Mikrobenschlacht", as one seasoned aquarist calls it – a microbial battle between the microscopical life forms that travelled to the tank on rock, stones, roots, plants, the sand and fine clay, the leaf litter, and even the glass and technology. The water turns milky. Whatever plants you have planted, they will disappear from sight. Suddenly, maybe even the next morning, the battle is over. The water has become crystal clear.

Constant filtration will make sure water remains clean, oxygenised, and clear – much clearer, indeed, than an actual Amazon river tributary would be. For an Amazonas river tribuary tank, dried oak and umbrella tree (Terminalia catappa) leaves will be added. Their humus matter helps softening the water, making it more acidic and suitable for blackwater fish and invertebrates – the swarming neon tetra and angelfish for example, that this chapter began with. The humus adds the kind of yellow, slightly brownish coloration typical for these waters that emerge deep from within the tropical rainforest.

Sumatra driftwood, roots of all sizes, is another staple of choice for all kinds of tanks, especially an Amazon one, as it too releases tannins into the water, turning it slightly yellow, more acidic, more soft. It is basic hardware you get at even the most modest pet store. It helps maintaining good water, and is a decorative choice as well: with it, you create zones, territories, layers. Places to hide below and watch over for other than human animals. Surfaces for plants and algae to cling to. Vistas for human onlookers, guiding their gaze, making the tank pleasant to look at.

Until you buy the first fish, up to three months will go by. The first ones, however, will be ones that are more forgiving of water less than perfect water. Multispecies-making, and this is true for the making of aquarium atmospheres as well, "carry forms of domination, communion, and activation" (Hayward 2010: 592) that are fraught with knowledge and responsibility. With a multitude of actors at play – leaves and bacteria, filtration, temperature, aquarists and water's chemical composition, (and the list could go on) making good water is skill that is not acquired without a great deal of experimentation at a cost. Aquariums-in-the-making very often do not become ecologically independent; they end up unable sustain themselves as balanced equilibriums. Beginner aquarists very often give up, leaving a trail of vanished subaquatic guinea pigs behind. Stories of failed design projects – stories of more-than-human lives vanished – speak of water gone bad.

Detailing the making: light, and life

The making of aquarium atmospheres, we have seen, is a complex weighing of aesthetic and ecological interest. It has been, historically, and so it is today. Lighting is another key pillar of those atmosphere-making practices. Light is more than "simply means of making things visible": from a making perspective, it is a key part of orchestrated social life that shapes sites and situations (Bille 2015: 259) and that enables conditions for perception (Böhme 2013). Light, through its many facets, tones, shadows, reveals both presence and absence; it is central to the everyday "practice of orchestrating spaces" (Bille 2015: 259). It shades the perception of all the other objects in an illuminated place. On the other hand, though, light, just like air and water, can be thought of as atmospheric matter itself. Something not to perceive, but to to perceive in, a phenomenon of lived experience (Ingold 2011: 96). And those lived experiences differ across species. Just like air and water, for visually-oriented animals and "sighted persons, [light] is the experience of inhabiting the world of the visible, and its qualities – of brilliance and shade, tint and colour, and saturation – are variations on this experience" (Ingold 2011: 128).

Let us move to a corner of the Waldhäusler aquarium store that we have not yet visited to illustrate this point further, and to see what we can learn from it about making aquarium atmosphres. In the silence of the store, the proximity of a tank is always implied by the murmuring sounds of the exterior filters behind the wood-paneled walls cleaning and turning the water. One such source of murmuring waters is connected to a seawater tank shining in an eerie, bluish light. It is here that I get a contemporary perspective on one of the most central pillars of aquarium atmospheres: light. Max has a new gadget. As he admits himself, he is not much of an early adopter when it comes to new technology. He rather likes to see it tested out in practice, over time, before he submits his aquariums and his fish and plants to it. LED (short for "light emitting diodes") illumination is such a new trend, and by now he has integrated it in his tank-making. LED is increasingly substituting the formerly used fluorescent tubes. And it can do many a thing, atmosphere-wise.

We are standing in front of a small saltwater tank right next to the entrance. As we stand, the light from Max's new LED strip shifts ever so slowly: from orange to blue, from greenish to reddish, through all the shades of the rainbow, at least to human perception. M ax draws my attention to a small mussel close to the pane. As the light goes brighter, the mussel slowly opens. Then it closes again, with the darker shades of blue. With the light, it moves, stuck to the ground, attentive to the ever-shifting qualities of its world. To many a Saturday shopper, the rainbow LED strip is a nice add-on, a new decorative element. But it is more, and it shows just how difficult it is to draw a line between ecological and aesthetic aspects of atmosphere in everyday multispecies life: light is needed by all species within the realm of the aquarium; last but not least, the humans depend on light so they can perceive the life beyond the glass. Usually, aquariums function independently of some aspects of larger weather patterns. They give up water into air, and air temperature and air pressure shape life inside the tank, yet sunlight, once a ressource early aquarist strongly depended on, tends to be avoided: its angle and intensity are difficult to control and may contribute to the growth of algae and over-heating. Where shoppers, however, see rainbow lights, a new possibility to enact the luminosity or dimness of their living room or commercial space (see Edensor 2017), a conch perceives light and darkness, possibility or danger, day and night. The whole biorhythm of the aquarium depends on humans choosing the right tone of lighting, flipping the switch in time or getting the setup of the timer switch right. Chemical water processes crucial for survival depend on light and dark; light is of course elementary for making aquatic plants grow and invertebrates thrive; 'night' and 'day', varying intensities of lighting and absence thereof, are needed so that plants can create the right chemical balance and aquariums can sustain themselves. Increasingly, light is being used to create even more differentiated moods. Programmable light-emitting diodes make it possible to create shadowy evenings and blue hours: orchestrated spaces that make it possible for even more lifeforms to experience light that suits their needs – or the needs of their owners. Aquatic forms of life, like any others, have complex and varied needs and preferences when it comes to light. The commercial light palette that sustains their possibilities of dwelling is adding more and more shades: dusk, dawn, twilight . . . The process of lighting up an aquarium illustrates how creating underwater atmosphere is a process of caring labour that requires technical and ecological knowledge as well as aesthetic skill and sensibilities. Hundreds of decisions like these coin the everyday creating and creative work of aquarium keepers, mediating between human and other than human needs and peculiarities.

Conclusion

The exploration of the ocean grounds at the turn of the 19th century set off an aquarium craze that triggered ecological experimentation, technological development, and aesthetic play with underwater worlds that one began dreaming about. 'Being underwater' became a romantically charged concept that superseded a former fear of the depth. Against this cultural and historical background, the aquarium, in its many shapes, became a site and situation of "ecological charisma" (Lorimer 2015): Historically, the very concept of 'ecology' and thinking about gaseous atmosphere as a life-sustaining envelope, have accompanied human tinkering with these 'coffee table oceans'.

The making of aquarium atmospheres by humans is worth sensing, tuning into, describing and thinking about as it caters towards two different kinds of audiences and atmospheric inhabitants, differentiated by their different sensoria and bodily needs: air- and soil-borne humans, and waterborne aquarium inhabitants. It adds fresh liquidity to the well-established atmospheric medium of air, and it sheds light on the part of more-than-human organisms in atmospheric making and becoming. Making "charismatic ecologies", relationally functioning microworlds fraught with affective appeal, from a human expert's perspective is an act of ongoing atmospheric sensitivity and translation in a world experienced through degrees of liquidity. Aquarium atmospheres are established through many complexly interwoven techniques – from making "good water" to the design of lighting to managing the coexistence of different species – catering to human and other-than-human needs alike and entwining their needs and wants. Aquarium onlookers take the air that they breathe when they enjoy the complexity of "affects, sensations, materialities, emotions and meanings [. . .] enrolled within the force-field of an atmosphere" (Edensor and Sumartojo 2015) for granted. For the lifeforms behind glass walls, even the physical atmosphere they live in is usually difficult to stabilise and their lives thus remain precarious.

Notes

1 A video recap of the 2012 Spring/Summer Ready-to-wear show offered by the official Chanel channel is available on Youtube, www.youtube.com/watch?v=xLw-hozWzpc (last accessed 7 June 2017).

2 Refer to www.oceanoptimism.com/ (last retrieved on 2 May 2016) for further information on the campaign. My sincere thanks go out to Elin Kelsey who invited me to be a participant in the event.

3 And owning an aquarium is almost synonymous with making it, as the tanks keep ever-evolving and need constant attention.

4 With important local adaptations: the Shanghai Ocean Aquarium, for example, I learned during the workshop, does not only represent local species, but has also been redesigned according to Feng Shui principles.

5 Cultivating an Amazon River Biotope Tank", retrieved 5 August 2016 from the website "Rate my Fishtank" a large online community for aquarists with more than 750,000 page views per month. www.ratemyfishtank.com/blog/cultivating-an-amazon-biotope-tank.

Bibliography

Anderson, B. and Ash, J. 2015. Atmospheric Methods. In *Nonrepresentational Methods: Re-Envisioning Research*, edited by P. Vannini. London: Routledge Publishing.

Bille, M. 2015. Hazy Worlds: Atmospheric Ontologies in Denmark. *Anthropological Theory* 15(3), 257–274.

Böhme, G. 2013. The Art of the Stage Set as a Paradigm for an Aesthetics of Atmospheres. *Ambiances*. Available at www.cresson.archi.fr/PUBLI/pubCOLLOQUE/AMB8-conf GBohme-eng.pdf [Accessed 14 July 2016]

Brunner, B. 2005. *The Ocean at Home: An Illustrated History of the Aquarium*. New York: Princeton Architectural.

Choy, T. and Zee, J. 2015. Condition-Suspension. *Cultural Anthropology* 30(2), 210–223. https://doi.org/10.14506/ca30.2.04

Edensor, T. 2017. *From Light to Dark: Daylight, Illumination, and Gloom*. Minnesota, MN: University of Minnesota Press.

Edensor, T. and Sumartojo, S. 2015. Designing Atmospheres: Introduction to Special Issue. *Visual Communication* 14(2), 251–266.

Griffero, T. 2015. *Atmospheres: Aesthetics of Emotional Spaces*. Farnham: Ashgate.

Harter, U. 2015. *Aquaria: In Kunst, Literatur, und Wissenschaft*. Heidelberg: Kehrer Verlag.

Hayward, E. 2010. Fingeryeyes: Impressions of Cup Corals. *Cultural Anthropology* 25(4): 577–599.

Hayward, E. 2012. Sensational Jellyfish: Aquarium Affects and the Matter of Immersion. *Differences* 23(3), 161–196.

Ingold, T. 2011. *Being Alive: Essays on Movement, Knowledge and Description*. London: Routledge.

Ingold, T. 2012. The Atmosphere. *Chiasmi International* 14, 75–87.

Ingold, T. 2015. *The Life of Lines*. London: Routledge.

Lorimer, J. 2015. *Wildlife in the Anthropocene: Conservation After Nature*. Minnesota, MN: University of Minnesota Press.

McCormack, D. 2008. Engineering Affective Atmospheres on the Moving Geographies of the 1897 Andrée Expedition. *Cultural Geographies* 15, 413–430.

Olalquiaga, C. 1998. *The Artificial Kingdom: A Treasury of the Kitsch Experience*. New York: Random House.

Pink, S. and Leder Mackley, K. 2016. Making, and Atmosphere: Routines of Home as Sites of Mundane Improvisation. *Mobilities* 11(2): 171–187.

Stewart, K. 2011. Atmospheric Attunements. *Environment and Planning D: Society and Space* 29(3), 445–453.

Wessely, C. 2013. Wässrige Milieus: Ökologische Perspektiven in Meeresbiologie und Aquarienkunde um 1900. *Berichte zur Wissenschaftsgeschichte* 36(2), 128–147.

World Association of Zoos and Aquariums (WAZA). 2009. *Biodiversity Is Life: An Educational Manual*. Bern: World Association of Zoos and Aquariums (WAZA).

8 Waves of experience

Atmosphere and *Leviathan*

Julia Bee and Gerko Egert

Coloured weather

An orange-reddish glow is bumping from the bottom into the blackness of the screen. Up and down. It pulsates rhythmically. Then it diffuses in the flurry movements of various colours that populate the first minutes of Lucien Castaing-Taylor's and Véréna Pavarel's 2012 film *Leviathan*. What emerges is a dance made of the boat's brownish-red glow amidst glistening flood lights at night, the greenish-white wave crests stirred up by the morning winds, the shimmering colours of the sea, reflecting and diffracting the darkness of the sky, and the flapping whites of the seagulls' flights.

The film is full of fleeting colours that in no way form clear-cut or even stable entities. They move in a constant flux, changing with the rising and setting sun, with its drying heat, with the wetness of the water, with the coldness of the wind. Processes of colouring: the whole scenario of the film is "colored by weather" (Taussig 2009: 251). One could even say that the colours themselves produce a meteorological scenario, a "colored weather" (ibid. 251) without any objects behind them: "[Y]ou cannot separate a color from what it is a color of,“ Michael Taussig writes (2009: 250). "Same as writing" (2009: 240), he subsequently adds. In this chapter, we will follow this concept of an inseparability of quality and entity, and transfer it to the movements of experience and atmosphere. Like the quality of colour and the process of writing, experience and atmosphere are not added to a pre-given thing or entity. Atmosphere, we argue, is a specific mode of becoming. It is the immanent relation that creates and combines the world as a non-linear chain of occasions.

Picturing the dazzling colours of the weather Taussig strongly emphasises how weather and aesthetic experience are entangled. Weather is full of experience without simply being its object. Coloured weather is atmospheric; it emphasises colour's processuality – colouring. He understands weather as an activity: In the interplay of its forceful movements, weather creates ever-new scenarios and colourful worlds.[1]

Weather is in all its colours not separable from any given material frame. It is part of the world's processes of becoming – its worlding.[2] Weather is – as Ingold (2011) describes – a scenario, without stable objects consisting of an ever-changing landscape. Like the seagulls in the sky, also the mussels, the pebbles, or the ripples in the beach are formations of the weather. There is no material landscape, no object previous to, what Ingold terms the "weather-world" (ibid. 126–135), only a meteorology of movement. These movements are not *of* the weather, but the weather *is* movement. "We are not required to believe that the wind is a being that blows, or that thunder is a being that claps. Rather, the wind *is* blowing, and the thunder *is* clapping. . . . " (ibid. 73). Standing with his students on a stormy day at the beach, Ingold describes the "weather-worlds" as follows: "We had [. . .] to recognise that the ground on which we stood was not really a supporting platform upon which things rest but a zone of formative and transformative processes set in train through the interplay of wind, water and stone, within a field of cosmic forces such as those responsible for the tides" (ibid. 131). He goes on to describe the movements of the sea and the birds:

> Against this background, we could dimly make out the wheeling forms of seabirds, but we recognised them not as objects that moved, but as movements. . . . We saw a world in movement, in flux and becoming, a world of ocean and sky, a weather-world. We saw a world without objects.
>
> (ibid. 131)

This complex interplay of forces, where one cannot differentiate between a given setting (landscape), a number of active players (wind, sun, seagulls, humans), and a set of actions (blowing, shining, flying, watching, moving) produces the weather-world.

Following the movements of weather, *Leviathan* becomes a *colour-weather-worlding*: the green-yellow-red dances of the mussels and starfishes under the sea, the silver-white-red of the fish splashing on the deck of the ship, or the strokes of the seagull's wings creating a choreography of contrasts in black and white. Even though one does not see any traditional weather scenarios – no oncoming storm or burning sun on the sea – *Leviathan* is full of weather. It is a weather of colours, of movements, and of experience.

Leviathan

The film *Leviathan* marks an important moment in the development of ethnographic research practices. The film addresses ethnography from a perspective of sensory experience. It adds to a long discussion on the relation of experience and abstraction as well as theory and sense in ethnographic research (e.g. Taussig 1993). In the *Sensory Ethnography Lab* in Harvard, where *Leviathan* was produced, sensory experience is actively involved on multiple levels. The studio's films (for example *Sweet Grass* [2009], *Foreign Parts* [2010], *Manakamana* [2013]) deploys a complex synesthetic approach to ethnographic filmmaking and therefore to ethnography itself by connecting vision, sound, movement and hapticality to a complex ensemble. Anthropologist Paul Stoller once coined the term "radical empirical anthropology" (1992: 213) in relation to Jean Rouch's anthropological films. "Radical empirical", we suggest, is an appropriate term for *Leviathan*, too, as will be shown below through the concept of radical empiricism by William James.

Leviathan follows the various movements of and around a fishing boat at sea of the New Bedford Coast. Weather thwarts any possibility of a linear narrative about a ship conquering the sea. With weather, one is kinetically, visually, and barometrically immersed in extreme scenarios. The movement of the sea and the boat are indistinguishable, and objects are often hard to recognise. Seasickness and nausea are caused by the lack of any stable frame of orientation or acoustic explanation.

Following the up and downs of the fishing vessel's movement, the camera dashes into the water. White spray floods the image. Every time the view moves above the water, one catches a glimpse into the black night and of the circling seagulls accompanying the boat's activities. The repetitious movements of the waves banging against the hull forge a rhythm of boat and sea, of sky and water. The bright green of the fishing net passes through the waters and onto the ship. The net hovers above the wet and shimmering deck, filled with fish waiting to fall into the machinic movements of washing, chopping, bleeding. One by one, the buckets of fish are brought below deck. A reddish-brown stream of waste and blood flows back into the sea. Seagulls dash into the water, diving for food. Tiredness and boredom run through the fishermen's movements, shaping the repetitious gesture of chopping off the fish heads and tails. Boredom dominates the operation of the crane, and the eyes staring into the blackness of the sea at night. And tiredness runs through every muscle of a saggy body fighting against sleep in front of the TV after a long day of work.

In the interplay of all these movements, some monotone, some flurry, some steady, some bursting the filmic atmosphere *Leviathan* emerges. These varying movements made the process of production quite complex. As co-director Castaing-Taylor describes the processes of filming: "We started off filming with good-ish professional HD cameras, but we lost them one by one to the sea, so the only cameras we ended up shooting with were these small digital SLR cameras and the tiny sports cameras called GoPros" (Castaing-Taylor 2014: 82).[3] The movement, the water, the always-unstable environment made conventional modes of filming impossible. In the film, GoPros are attached to the bodies or the heads of the fishermen, but also to the ropes, the anchor and the crane. The cameras swim with the fishes in the water; they follow the seagulls, and they swash back and forth with the waste in the basins on board the ship. These cameras do not operate from a distance; they capture the moving meteorology from within. They work like barometers, sensing the changes of their surrounding by capturing the multiple movements and flows. Their super wide-angle shots (captured by fisheye lenses) and the extreme close-ups produce intensive images. Together with "an acoustic ecology, produced by new machinic agents" (Kara and Thain 2014: 193) they create "an immersive, materialist holding together of the film's elements without a teleological arc" (Kara and Thain 2014: 194).[4]

Using multiple cameras and heterogenous perspectives, the film is not restricted to the human perspective. The cameras move round, they go up and down with the waves; attached to the ropes or to long sticks they move across the ship. They do not stop at the guardrails, but cross through the water and the air, moving with the swarms of seagulls and dead fishes. *Leviathan* refuses any central perspective: it does not represent a single human perspective (nor does it represent the perspective of *the* fish, *the* seagull, *the* ship, etc.). In *Leviathan,* the camera becomes part of these oceanic, biological, animalistic, technological dances. The various and intersecting movements and perspectives of the different cameras compose the meteorological choreography of the movie.

Immanent atmosphere

How do *Leviathan*'s movements relate? We suggest that its movements are gathered by atmosphere, whilst at the same time the film's atmosphere is also co-created through the way these movements relate. Atmosphere is nothing secondary to any given situation. It does not envelop an object (the ship, the fish, the fishermen) or a movement (the swimming, the fishing). With its multiple cameras, the film captures the atmosphere barometrically. *Leviathan* starts in the midst, and unfolds by taking up and recomposing various movements in the acts of filming, editing and watching the film. Atmosphere becomes the very specific mode of how *Leviathan* unfolds. Alfred North Whitehead calls this unfolding the "affective tone" (Whitehead 1967: 180).[5] The affective tone is at the heart of Whitehead's concept of the event: it is the way the event unfolds. By feeding into each other, the occasions relate and form the movement affectively and atmospherically: atmospheric movement events. All of the movement in *Leviathan* is made of occasions feeding into each other, forming a chain. Chaining thus becomes movement's atmosphere.

Leviathan does not consist of discrete shots or movements. It is not a sequence of disparate chain links or elements, like the wavering ship, the sea, the rotating crane or the tired fishermen, but is a sequence of events. The movements in *Leviathan* cut across the various shots and compose a series of events that feed into each other. The relations between the events are not a linear effect and cause. It is the event's affective tonality taken up by another event. It is a qualitative "carry-over". As Brian Massumi explains: What the affective tonality "does is carry-across the qualitative nature of what happens. It gives an abstract, purely qualitative background continuity to the two moments" (Massumi 2011: 65). The affective tonality of one event becomes the first phase of another one. Atmosphere is the immanent relation that creates and combines movements as non-linear chains of occasions.

These movements do not form any kind of coherent entity. With every event they deflect their direction and shift the whole choreography. There is no atmosphere of *one* movement (this could be called the movement's affective dynamic, its rhythm, its speed or force). Rather, atmosphere is the way movements relate and choreograph the intensive milieu – it is the way affects compose (and thereby we include human as well as nonhuman affects). This filmic choreography is atmospheric in a double sense: at one and the same time, atmosphere is the movement's immanent relation, the becoming of its continuity, *and* the process of differentiation of the atmospheric milieu. In atmosphere relation and difference are no opposites but mutually inclusive. With every linkage of movements and every production of continuity new differences emerge. In atmosphere, there is no division between act and milieu – both are folded into each other. Act and milieu co-evolve simultaneously in a field of differential becoming.

As a force of qualitative carry-over, atmosphere does not exclude content. Often, content is seen as the already given in a particular situation. It is the ground that creates atmosphere as a second level. Content, in case of *Leviathan*, works the other way round: "[t]he contents are precipitation" (Massumi 2011: 66). The fishermen's work, the dead fish, or seagulls flying create a "rain of [. . .] gestures in the micro-climate that is life at this moment, coming in drops" (Massumi 2011: 66). This content is part of the atmosphere's specific mode of actualisation. In this process of formation, new movements, new differences and new affective tonalities emerge. The atmospheric force does not dissolve in content. Like every wave flows back into the sea of movements, every actual gesture, image or thing also feeds back into the process of atmospheric becoming.

The atmospheric choreography of *Leviathan* is a composition of differential becomings. Atmosphere is the shape-shifting force of its own alteration. Only in its alteration can atmosphere be perceived. The slowness and speed of the movements, the rhythms of the waves or the (sonic) refrains of the crane, the darkness of the sea and the sky, the flickering white of the seagulls, the brownish-red of the boat, and most of all, the contrast between these colourful movements are what compose the atmospheric choreography called *Leviathan*. The film's atmosphere is in no way coherent. It is an ongoing process differentiating itself into multiple – often contradictory – intensities: hectic pace, tiredness, threatening darkness, and the repetitive and calming sound of the waves. None of these tonalities is in themselves stable. They are, instead, a precarious meta-stability brimming with potential to become different – to alter its speeds, its intensities, its brightness and colours, its affective tonality.

Waves of experience

Experience is key in atmosphere's differential becoming. Like atmosphere, experience is not added to a given scenario. It is part of the process of atmosphere's unfolding. As *Leviathan* constantly experiments with points of view from material and nonhuman perspectives, experience cannot be attributed to the human subject as its centre. The perspectives in the film are neither human nor do they belong to specific nonhumans: instead, the result is an atmospheric experience consisting of many perspectives from within a certain environment or milieu. One could even say *Leviathan* is one such milieu. Moreover, it is an atmospheric milieu, as it is not strictly divided into material and immaterial parts. This multi-differentiated human and nonhuman material and immaterial characterisation of experience can also be found in the work of philosopher and psychologist William James (1912). He writes about experience, comparing it to the rolling up and down of waves, rather than as an objective account of events:

> We live, as it were, upon the front edge of an advancing wave-crest, and our sense of a determinate direction in falling forward is all we cover of the future of our path. It is as if a differential quotient should be conscious and treat itself as an adequate substitute for a traced-out curve. Our experience, inter alia, is of variation of rate and direction, and lives in these transitions more than in the journey's end.
>
> (ibid. 69)

Here, the wave is an experience itself: a non-personal feeling that consists in its "speeds and slownesses" (Deleuze and Guattari 1987: 262) of rolling and vanishing. In James' philosophy of so-called "radical empiricism" (James 1912), experience is chance, flux, stream and becoming. It refuses a stable representation in the subject of events taking place outside.[6] Experience is an immanent event of chance in nature and culture, in waves, humans, in seagulls, as well as their flight and the waves' movements. It neither represents nor consists in one object, one motive, or one isolated event taking place; instead, it is made up of the constant interweaving of multiple processes and events.[7] Experience in *Leviathan* is an atmospheric, immersive and a non-reliable source for positivist thinking, even though it is a documentary film. Nonetheless, it takes up and creates an atmospheric event by focusing on relations and processes and not on isolated objective accounts of reality. Although James is a philosopher and *Leviathan* is a documentary, they both share a concept of experience that is different than the positivist idea of experienced-based research practices. Their shared understanding is a sensational, sensory and atmospheric construction from within an atmosphere, and not from an objective distance (see also Thain 2015). Their thinking can be characterised as a movement within the *flux* and the waves of experience.

Experiential atmosphere

The concept of the sensory, as in the name of *Sensory Ethnography Lab*, can be related to the notion of atmosphere. Atmosphere is a complex of material and immaterial "tissue[s]" of experience that constantly "grow[s] by its edges" (James 1912: 87), meaning it constantly differentiates and weaves new ends. James himself states that "thoughts in the concrete are of the same stuff as things are" (1912: 37). Deriving from James' thought and from *Leviathan*, atmosphere can be understood as the intermingling, the relation and the process of becoming of experience. For James, "experience as a whole is a process in time" (1912: 62). It is a woven tissue that consists not in subjects that perceive a world: it is the very change the world consists of. A world he termed "a world of pure experience" (1912: 39ff.). "Experience itself, taken at large, can grow by its edges. That one moment it proliferates into the next by transitions which, whether conjunctive or disjunctive, continue the experiential tissue" (James 1912: 87). Here, experience is neither objectivist nor subjectivist (see Massumi 2011: 29ff.). On the contrary, in James' notion of experience "pure experience" (1912: 39–91) is divided into subject and object afterwards: " 'inner' and 'outer' are not coefficients with which experiences come to us aboriginally stamped but are results of a later classification performed by us for particular needs" (1912: 146). In his philosophy, he crafts a *world of (pure) experience.*[8]

In the atmosphere of thoughts articulated in and with *Leviathan* and James, one can think about the medium film and its connection to atmosphere neither as an unmediated, purely documentary notion of reality nor as a simple opposition to subjective perspective. On the contrary, *Leviathan* becomes the modulation, construction, and crafting of the reality: Film is not the medium of a reality given to the camera and the sound recording. Rather, *Leviathan* is itself a process of change and flux, and it takes part in the construction of the atmosphere. Film itself, especially and intensely *Leviathan*, is crafted from experiences. It actively takes part in the differentiation of the *milieu of experience*. It becomes a process of taking up an atmosphere and constructing a different process that exceeds what is given and produces something new. Differentiation of the atmosphere thus has productive as well as receptive aspects.

Radical atmospheric filmmaking

Leviathan concentrates and intensifies concepts of experience and sensation in a pictorial or audio-visual way that is key for ethnographic film in general. In its modes of documentation, its perspective is from within the atmospheric milieu and without a verbal commentary framing the experience. It is an experience-centred way of doing visual research. By using images, sounds, and movements as a trans-sensual form of media, it stimulates the spectators' visual or auditory as well as kinaesthetic senses. As a film, *Leviathan* is felt and it is a philosophy of experience at the same time. Yet it is not *about* experience. Rather, it consists *in* experiences that are of a complex intermingling atmosphere. This experience is not only human, it is "more than human" (Manning 2013: 89).

Lucien Castaing-Taylor once used the reference to philosophies of experience himself as a framework for doing ethnographic film: In an interview with ethnographic filmmakers Judith and David MacDougall, Taylor terms the style of filmmaking close *to* and from *within* an experience a "radical documentary filmmaking" (Barbash and Taylor 1996; Taylor 1996; in relation to Rouch's work: Stoller 1992).[9] Here he refers to James' concept of "radical empiricism". Years later, it is *Leviathan* that takes up James philosophy of "flux" and "stream". It seems as if Taylor, when writing about research practices of ethnographic film, also references James implicitly (Taylor 1996). He uses film, as opposed to discourse and text, as a medium that focuses on the complex relations, movements and changes that happen as ethnographic research. The atmosphere of the film can be understood exactly as the running together of material and immaterial forces, things, events and affects that specifically intermingle in the event of producing an ethnographic film. In his films, Castaing-Taylor (with Ilisa Barbash in *Sweet Grass* and Verena Paravel in *Leviathan*) takes up the notion of experience he found in James and develops it further on another plane of thought. In his 1996 text, Taylor is concerned with discussions in anthropology and ethnography about film, in which the medium is being accused of not having sufficient distance to be considered scientific. In 2012, he and Verena Paravel affirmed the constructivist, modulating and synesthetic way of experiencing the complex environment or milieu (not only a social milieu but one of things, people, movement and affects) in their radical empirical way of filmmaking.[10]

The film is a process built from the atmosphere of the milieu of the New Bedford Cost. It is a form of *reception* of the atmosphere of which *Leviathan* consists. The complex interweaving of processes like fishing, the works on deck of the ship, movements of the workers, the flights of the seagulls, and the constant movement of the ship build an atmosphere that is taken up by the various cameras' perspectives. The camera and its point of view are not secondary to the existing relations and processes, but they produce and further differentiate the milieu of what is happening around the fishing boat. Experience is also another becoming or *production*. The film evaporates from the complex milieu as (nonhuman) experience: The film composes and is composed with and by perception. It is a "weaving of experience" as James describes the becoming of the world, a world that consists of experiences that are material and immaterial at the same time.

Here, the spectator's experience might be understood as taking part in the modulation of the atmosphere of the film and not as perceiving it from the outside or as a secondary force. Atmosphere might be understood as taking part in the constant play of documentary as well as in the construction of artful images – an art that cannot be reduced to the work of a human artist. Here, experience composes the film as well as the spectator: Inside and outside (subject and object) are not the proper starting points to describe experience in *Leviathan* from a radical empiricist perspective.[11] Just as the camera does move with the milieu around the fishing boat, the spectator's perception is also interwoven in these processes. It is actively forming a new atmospheric assemblage of experience.

The construction of *Leviathan* is not a human one. It is the complex doing of an atmosphere *as* reception and production at the very same time. The atmosphere is another form of abstraction; something that cannot be reduced to the accumulation of the doings of people or even the interplay of cultural, social or natural forms of meaning. Abstraction is a construction that works and modulates atmosphere as sensual perception and affective process. Not only do subjects perceive atmosphere, atmosphere is an experience *as* abstraction.

Here, atmosphere is neither the glue that connects the subject of the research to its object; nor is it a medium, in the sense that it is the secondary relation between two already given entities. Atmosphere is at the heart of the events unfolding. Like the affective tonality, discussed by Whitehead, it is "something we find in ourselves, not something we find ourselves in" (Massumi 2011: 65). In the atmospheric event, knowledge emerges as a differential becoming. Experience is therefore not an experience *of* the atmosphere but is an *atmospheric experience*: it neither starts nor ends with a secured set of facts, it creates a shape-shifting choreography of manifold movements.

Atmosphere and media are not necessarily situated on different planes, in a sense that media apparatuses *represent* experience only. This is an argument often used to resist naïve realism. Creativity, construction and nonhuman experience and the plane of experience as reception-production are in no way contradictory. The aesthetic crafting of experience as a taking-up (reception-production) of atmosphere, i.e. the relational processes of entities, affects, things and thoughts create a field of pure experience as the Jamesian concept describes it. Therefore, the medium of film is in the midst of the events around the fishing boat – it is part of the atmospheric events of experience as an emerging form of film, of sound and of movement.

Speaking with James, the *stream of consciousness* is neither pre-mediated nor mediated, but *immediate* in the way that the medium emerges as a production of atmosphere. This might be another sense of sensory ethnography of atmosphere. Here, the connection of atmosphere as the very material, the tissue of experience, can contribute to discussions around ethnographic filmmaking. However, atmosphere is seen not as something that resists depiction or loses reality in its mediation. Following James and *Leviathan*, one can affirm a nonhuman, atmospheric, constructionist style that operates *in* and *with* the milieu of the New Bedford Coast. It takes part in a procedural and relational approach to further build up the atmosphere.

Like James' wave crest, the film consists of experiences. It is made of movements and perceptions that evolve around the boat, the shipping, the working, the colours, etc. The film is part of the sensation of the events and *how they enrol in time*. It is a form that takes part in and mediates the atmosphere without only giving objective access to what is happening. Atmosphere can be seen as a productive force that builds a film. It is what we see *in* the film as well as the processes of cutting, editing, etc. Here, atmosphere provokes new experiences as ongoing process: the milieu of experience "can grow by its edges" (James 1912: 87).

Leviathan's perspectivism

Filming is the production of perspectives – as is ethnography (Viveiros de Castro 2014). In ethnography, perspective is often discussed as either objective or subjective. *Leviathan* offers a new perspective on perspectives. It was the emphasis on the perspective that led to the acknowledgement of the filmmaker's point of view in ethnography. *Leviathan* puts this more radically and turns towards the perspective as *relation*. As atmosphere is not the atmosphere *of* something (e.g. an ensemble of a certain number of material things or processes), the filmic perspective is not directed at a given object. The film rather uses perspective to create new relations. The cameras differentiate the process of fishing and turning it into the filmic event *Leviathan*.

In the process of adding and multiplying the perspectives in and of *Leviathan*, the film creates other ways of fishing. These perspectives are not partial in a sense that only their sum can get us the whole picture. The proliferation of perspectives creates a world consisting of nothing but perspective: perspectives in their transversal differentiation.

Multiplying the perspective does not lead to a relativism of experience, but to perspectivism. If relativism argues for multiple perspectives on a pre-given world, perspectivism – as Eduardo Vivieros de Castro argues – calls for a world that is made of multiple perspectives.[12] Perspective changes the world *as such*.[13] There is no object, or world, not even process or movement behind the experience and its perspective. Perspective becomes itself processual. It is the perspective that worlds. This process of worlding is composed of the sea, the fishing, the ship – and here especially – the atmosphere *Leviathan* is made of. In its proliferations of perspective, *Leviathan* differentiates the given: Every perspective opens up another possible world and thereby questions the very modes of existence of the existing world. Following Deleuze on the notion of the possible world, "the possible is not here an abstract category designating something which does not exist: the expressed possible world certainly exists, but it does not exist (actually) outside of that which expresses it" (Deleuze 1990: 307).

Vivieros de Castro's perspectivism shifts the goal of anthropology from the question of epistemology to the question of ontology (or, to be more precise: to ontogenesis). *Leviathan* shifts the perspective of experience from a phenomenological to a radically empirical documentary. By multiplying the perspectives of experience, the film creates many possible worldings.[14] This is fishing as fishing was never before.

In the play of numberless perspectives, atmosphere is experienced as manifold. Here, atmosphere is not *of* something – e.g. a perspective – but a multiplicity of atmospheric becomings. In the same way the act of worlding does not describe the creation of an individual world; perspective is not the core of an individual atmosphere. The creation of perspectives is rather the ongoing process of atmosphere differentiating itself. This is not about *Leviathan*'s atmosphere but about *Leviathan*'s opening up of new differentials in the atmosphere immanent to its processual becoming. Only when atmosphere is infinite can one think of it as a modulation or mode of experience, which is not attributed to anything pre-given (be it a subject or object, a specific culture, an object of investigation, a practice, a process, a movement or a perspective).[15] In its immanent infinity, atmosphere is not enveloping anything but differentiates in ever-new ways. In these atmospheric differentiations the human and nonhuman perspectives of *Leviathan* produce new experiences *as* possible worlds.

Conclusion – or a passage in the atmospheric sea of sensations

Drawing from various process-oriented theories we regard atmosphere as a concept that moves beyond dichotomies of subject and object in film and ethnography likewise. The atmospheric force does not envelop any content. It is relational yet contrary to the idea of a fog that glues together pre-given entities of subject and object. Atmosphere is a milieu that consists of manifold heterogeneous perspectives and therefore overcomes the strict division of material and immaterial qualities. Thereby it crosses any division between act and milieu – in atmosphere both are folded into each other. The atmospheric perspectivism proposed in this paper embraces a productive excess of experiences. Experience – in the sense of James' radical empiricism – is not a representation of atmosphere in the receiver. It is not even a perspective on atmosphere from outside. Rather it is immediately participating in atmosphere's proliferation of perspectives: Experience is *of* atmosphere.

Leviathan expresses how atmosphere operates immediately as process of becoming. As an immanent force atmosphere modulates the way experience unfolds. Experience is at the very heart of atmospheric processes: Like experience, atmosphere is a flux of (dis)continuous differentiation, creating ever-new differences. Yet, in atmosphere, relation and difference are no opposites but both part of the very same event. *Leviathan*'s excess of (non)human perspectives demonstrates how experience contributes to the atmospheric milieu from within. Perspectives create *Leviathan* like the sea consists of waves and in the multiplicity of waves ever-new atmospheric patterns emerge. *Leviathan* – a passage in the atmospheric sea of sensations.

Notes

1 In the last chapter of his book *The Color of the Sacred*, Taussig turns towards the relationship between color and weather. By borrowing Marcel Proust's phrase "colored by weather", (2009: 251) he focuses on the processuality of colouring. Turning towards Nietzsche, he emphasises the movements and different speeds of this "colored weather" (2009: 252).

2 With the concept of worlding we foreground the world in its flux of becoming. The processes of worlding are manifold and neither start nor end with "the world". Erin Manning writes: "An otherness of worlding is always more than one. It composes-with experience, refuting the notion that the world is already known, pre-formed. This worlding is thought in motion, thought individuating in an amplifying incorporeality, a vibratory materiality" (Manning 2013: 169).

3 GoPro cameras are used most often to film action sports like skiing and surfing. Normally these cameras are attached to a helmet or body, but they can also be mounted on various devices. In her contribution to a special issue of the *Visual Anthropology Review* (2015) entirely dedicated to *Leviathan*, Alanna Thain describes the excessive use of GoPro cameras as a turn toward an "observation without distance" (2015: 47). Thereby she connects the filmic practices to Raymond Ruyer's concept of *survoler*: "Survoler, here as a tactic of sensory ethnography, refuses the corrective distancing from sensation as a way of knowing the world, proposing an immanent alternative to a politics of representation through ethico-aesthetic experience". (2015: 42).

4 Here, the notion of ecology is – like weather – not limited to nature. The "eco-logic" as proposed by Félix Guattari (2000: 44) rather describes the linkages of such diverse realms such as the social, the psychic, the economic, the physical, the acoustic or the visual.

5 In *The Adventures of Ideas*, Alfred North Whitehead develops the concept of "affective tone" involved in the occasion of experience: "It must be distinctly understood that no prehension, even of bare sense, can be divested of its affective tone, that is to say, of its character of a 'concern' in the Quaker sense. Concernedness is of the essence of perception" (Whitehead 1967: 180).

6 "To be radical, an empiricism must neither admit into its constructions any element that is not directly experienced, nor exclude from them any element that is directly experienced. For such a philosophy, *the relations that connect experience must themselves be accounted for real as anything else in the system*" (James 1912: 42).

7 His central argument about experience is its relational character expressed in formulations like the following: "Every examiner of the sensible life *in concreto* must see that relations of every sort, of time, space, difference, likeness, change, rate, cause, or what not, are just as integral members of the sensational flux as terms are, and that conjunctive relations are just as true members of the flux as disjunctive relations are. This is what in some recent writings of mine I have called the 'radically empiricist' doctrine (in distinction from the doctrine of mental atoms which the name empiricism so often suggests). Intellectualistic critics of sensation insist that sensations are *disjoined* only. Radical empiricism insists that conjunctions between them are just as immediately given as disjunctions are, and that relations, whether disjunctive or conjunctive, are in their original sensible givenness just as fleeting and momentary (in Green's words), and just as 'particular,' as terms are" (James 1912: 4).

8 Literary artists like Virginia Woolf and Marcel Proust experimented with the notion of *stream of thought* or *stream of consciousness* – two concepts that James addressed in his *Psychology* from 1892 (James 2001).

9 Scott MacDonald (2013) argues that experience as a whole is key for Harvard School of ethnographic film. He also sees a strong influence of the philosophy of William James in the works of Robert Gardner, Timothy Ash and Lucien Castaing-Taylor.

10 Lucien Taylor and Ilisa Barbash used the term "radical empirical documentary" to characterise MacDougalls Style of filmmaking (Barbash and Taylor 1996).

11 The division of subject and object, knower and known comes afterwards for James: "Knowledge of sensible realities thus comes to life inside the tissue of experience. It is *made*; and made by relations that enrol themselves in time. Whenever certain intermediaries are given, such that, as they develop towards their terminus, there is experience from point to point of one direction followed, and finally of one process fulfilled, the result is that their starting point becomes a knower and their terminus an object meant or known. That is all that knowing (in the simple case considered) can be known as, that is the whole of its nature put into experiential terms" (James 1912: 57).

12 In his discussion of Amerindian thought Vivieros de Castro exposes the concepts of perspectivism and multinaturalism: "(Multi)cultural relativism supposes a diversity of subjective and partial representations each striving to grasp an external and unified nature, which remains perfectly indifferent to those representations. Amerindian thought proposes the opposite: a representational or phenomenological unity which is purely pronominal or deictic, indifferently applied to a radically objective diversity. One single 'culture', multiple 'natures' – perspectivism is multinaturalist for a perspective is not a representation". (Vivieros de Castro 1998: 478)

13 Viveiros de Castro phrases this idea in the pointed sentence: "[A]ll beings see ('represent') the world *in the same way*; what changes is *the world they see*" (Viveiros de Castro 2014: 71).

14 In his discussion of Michel Tournier's novel *Friday*, Gilles Deleuze describes the concept of the other as "the expression of a possible world" (Deleuze 1990: 308). "But the Other is neither an object in the field of my perception nor a subject who perceives me: the Other is initially a structure of the perceptual field, without which the entire field could not function as it does" (Deleuze 1990: 307).

15 A similar argument can be found in Viveiros de Castro's discussion of nature: "[H]uman nature could be conceived as something like a minimum common multiple of difference – bigger than cultures, rather than smaller – or something like the partial integer of the different relational configurations we call 'cultures.' The 'minimum,' in this case, is the multiplicity that is common to humans – humanitas multiplex. Thus conceived, nature would no longer be a self-same substance situated within some

naturally privileged place (such as the brain, for example)" (Vivieros de Castro 2013: 481–482). This question of a multiplying nature, a multinaturalism, is also very closely connected to his writings on anthropology's turn towards ontology (see Viveiros de Castro 2013, 2014).

Bibliography

Barbash, I. and Taylor, L. 1996. Radically Empirical Documentary, an Interview With Judith and David MacDougall. *American Anthropologist Journal* 98(2), 371–387.

Castaing-Taylor, L. 2014. I'm a Geographer: Interviewed by Olivia Edward. *Geographical* 86(1), 82.

Deleuze, G. 1990. *The Logic of Sense*. New York: Columbia University Press.

Deleuze, G. and Guattari, F. 1987. *A Thousand Plateaus: Capitalism and Schizophrenia*. Minneapolis, MN and London: University of Minneapolis Press.

Guattari, F. 2000. *The Three Ecologies*. London and New Brunswick, NJ: Athlone Press.

Ingold, T. 2011. *Being Alive: Essays on Movement, Knowledge and Description*. London and New York: Routledge.

James, W. 1912. *Essays in Radical Empiricism*. New York: Longmans, Green and Co.

James, W. 2001. *Psychology*. New York: Dover Publications.

MacDonald, S. 2013. *American Ethnographic Film and Personal Documentary*. Berkeley, CA: University of California Press.

Manning, E. 2013. *Always More Than One. Individuation's Dance*. Durham, NC: Duke University Press.

Massumi, B. 2011. *Semblance and Event: Activist Philosophy and the Occurent Arts*. Durham, NC: Duke University Press.

Stoller, P. 1992. *The Cinematic Griot: The Ethnography of Jean Rouch*. Chicago and London: Chicago University Press.

Taussig, M. 1993. *Mimesis and Alterity: A Particular History of the Senses*. London and New York: Routledge.

Taussig, M. 2009. *What Color Is the Sacred?* Chicago and London: The University of Chicago Press.

Taylor, L. 1996. Iconophobia: How Anthropology Lost It at the Movies. *Transition* 69, 64–88.

Thain, A. 2015. A Bird's-Eye View of Leviathan. *Visual Anthropology Review* 31(1), 41–48.

Thain, A. and Kara, S. 2014. Sonic Ethnographies. Leviathan and New Materialisms in Documentary. In *Music and Sound in Documentary Film*, edited by H. Rogers. New York: Routledge, 186–198.

Viveiros de Castro, E. 1998. Cosmological Deixis and Amerindian Perspectivism. *Journal of the Royal Anthropological Institute* 4(3), 469–488.

Viveiros de Castro, E. 2013. The Relative Native. *HAU: Journal of Ethnographic Theory* 3(3), 473–502.

Viveiros de Castro, E. 2014. *Cannibal Metaphysics: For a Poststructural Anthropology*. Minneapolis, MN: Univocal.

Whitehead, A. N. 1967. *Adventures of Ideas*. New York: Free Press.

9 From affective encounters to wearable forms

Fashion design pedagogy and the creation of atmosphere

Todd E. Nicewonger

Introduction

It was late spring when I went to visit with Isabella, a fourth-year student at a fashion design school in Antwerp, Belgium, where I had been conducting fieldwork.[1] We had first met seven months earlier when I watched her cover a large swath of a classroom floor with a menagerie of images: celestial maps, pictures of outer space and photos of African cultural activities and objects. Curious to find out more about her project I asked her what the images were for. She explained that they were the sources of inspiration that she was hoping to base her collection on.

Fascinated by the eclectic nature of the imagery, while also wanting to better understand how the images figured into her design process, I closely followed Isabella's progress in the coming months as well as her classmates. Like Isabella, the other students at the design school spent the majority of the first part of the fall semester conducting extensive research for their design collections. This research took on many material and aesthetic forms and while the source materials of students' collections varied, their efforts converged around a shared goal: to identify research materials that reflected the "feeling" or "atmosphere" of the particular kind of design collection they sought to create.

Consequently, it was not surprising that as I sat in Isabella's kitchen on that late spring afternoon, I found myself once again looking at many of the same images I had observed her working with at the start of the school year. But instead of being neatly spread across the floor of a classroom, the images (along with many others she had collected over the past few months) now filled several artfully crafted, multi-paged design portfolios.[2] The pages included hand-written reflections, pictures of material experiments and elaborate collages depicting richly illustrated scenes of an imaginary world. Like many of the students at the academy, this world, or atmosphere, had been inspired by a socially significant issue, which Isabella deeply cared about.

In Isabella's particular case, the idea for her collection had originated prior to the beginning of the school year when she had been following news reports describing the plight of African migrants who were trying to cross into southern Europe. Deeply alarmed by the horrific ordeals that these migrants were enduring, she began searching for ways of reflecting on these events through her design work. In the process she drew on nautical and celestial imagery to reference the harrowing journey that the migrants faced as they attempted to cross the Mediterranean Sea in search of economic and social opportunities. Additionally, she explored ways of speculating about how these issues might play out in a different time and place by drawing on images depicting space travel and various astronomical symbols, which she integrated with pictures of African social life to create a wholly new world: an atmosphere for her design collection. As a result of this research these resources became the conceptual groundwork of her collection.[3]

My interaction with Isabella that afternoon was not unique. During my fieldwork, design students constantly evoked atmosphere as an idiom to describe the affective force of aesthetic figures and images. This was no coincidence: atmosphere also played a key role in classroom instruction. I discovered that teachers relied on a rubric composed of several key terms – including "atmosphere" – to evaluate the efficacy of student design work.[4] As such, this category was central to the creation of a collective interpretive schema by which teachers and students alike were able to articulate the qualities of abstract designs that otherwise could seem whimsical, idiosyncratic or simply unintelligible. Through the institutionalisation of this concept, aspiring designers learned how to draw on moral and social issues as inspiration for experimenting with various bodily forms, methods and design imaginaries (see Nicewonger 2011). In what follows, however, I am interested in their expansive affective possibilities. That is, I will argue that the category of atmosphere is best understood as a mode of practice through which students learn to re-interpret the causal relationships between aesthetic forms and social imaginaries.

To be clear then, I am not arguing that students like Isabella necessarily see their designs as making a direct impact on the social issues that inspire their fashion collections. Instead, I am arguing that through the research practices that these students learn to conduct in order to define the atmosphere of their design collections, they are able to manage the conceptual gap that stands between the social imaginaries that are inspiring their design work and the actual materialising practices that go into the making of wearable design forms. What this reveals is how wider cultural symbols and affective idioms come into being and are distributed through the practice of fashion design. In this way, we can begin to see how Isabella and her fellow students are not just being taught to draw on the concept of atmosphere to produce fashion forms that interpret popular trends in the clothing market. Rather they are learning to create conceptual designs that allow both viewer and wearer to inhabit imagined worlds. This means that they are engaging in a particular kind of fashion design practice that centres on the development of materialised concepts or what the students at the academy described as the creation of an imaginary world (Clark 2012).

Subsequently, this chapter bridges a number of important projects in contemporary ethnography by bringing together work on affect (Hayward 2012; Chumley 2013; Stewart 2007) and world-making (Haraway 2011; King 2011; Tsing 2015). In the past two decades anthropological attention to the everyday affects of worlding on lived experience have grown in tandem with the field's engagements with affective theory, which itself reflects a move away from "social constructivist" framings of culture (e.g. social constructionism, post-structuralism, see Rutherford 2016: 287–288). As a result, ethnographers are thinking "hard about the ways that passions pass between bodies. It is inviting them to see in the things people say, write, make, and do together the effects of movements that breach the boundary long presumed to exist between passion and reason, body and soul" (Rutherford 2016: 287). Kathleen Stewart's work on atmosphere develops this point further. Repurposing Heidegger's notion of worlding – "an intimate, compositional process of dwelling in spaces that bears, gestures, gestates, worlds" (2011: 445) – as an analytic framework for locating the social within the creative capacities of subjects, Stewart shows how worlding is not experienced through representational forms. Rather it becomes knowable through the "qualities, rhythms, forces, relations, and movements" that impinge on one's interactions with the material-social world (ibid.). These qualities or forces are born out of affective perceptions no less than narratives that invoke the intensity of being part of a shared "something" as that "something [is] happening" (Stewart 2011: 447). In this way, the affects of atmosphere may form through the glimmer of a fast-approaching headlight or the music-like creaking of a rusty hinge. In either case, these events and encounters with the material-social world impress upon people's senses in ways that can give rise to new ways of "living through things" (Stewart 2011: 452).

The following essay explores how people in an educational institution come to be collectively attuned to such atmospheric sensations of something social that are not yet formed or a "structure of feeling" that is still "in solution" (Williams 1977: 133). To do so, I examine how students at the academy were taught to define the atmosphere of their fashion collections. Attending to these processes requires not just acknowledging that designers draw on affective experiences as sources of inspiration for their design work. Rather it also involves examining the ways in which these practices are patterned and enriched with cultural symbolism. Design means generating social contexts for making, talking about, and evaluating how affective and material resources animate creative processes (Murphy 2015; Ingold 2013). Such practices, I argue, provide important insights for developing multimodal modeling practices that can be taken seriously by and incorporated into ethnographic research. This account begins with a brief description of the school's pedagogical history and teaching philosophy in order to further situate these practices within the academy's wider curricular aims.

Antwerp and fashion

When writing about Antwerp fashion, it's almost impossible not to reference the story of the Antwerp Six, since the two are so intimately intertwined. The Antwerp Six are six fashion designers who in the late 1970s, early 1980s studied in Antwerp, Belgium (Van Godtsenhoven 2013). After completing their studies this group, which included Walter van Beirendonck, Dries van Noten, Dirk van Saene, Dirk Bikkembergs, Ann Demeulemeester and Marina Yees, went on to gain international notoriety for their experimental style and innovative design work, which in turn helped establish Antwerp as a capital of high fashion.[5]

Today the story of the Antwerp Six continues to be closely associated with Antwerp's contemporary fashion scene (e.g. Menks 2013). But the city has also become widely celebrated for its innovative approach to fashion education. In fact when I arrived in Belgium in the mid-2000s I met students from all over the world who had come to Antwerp to study fashion design, inspired in part by the idea of being able to carry out their schoolwork in the same city that Antwerp Six had studied.[6] These students also associated the school's teaching style with the work of such conceptually driven designers as Jean Paul Gaultier, Yohji Yamamoto and Vivienne Westwood who the students often described as having an "avant-garde" design style, which is also how they characterised the school's curriculum.[7] "Its more art-like here", was how several students explained the school's pedagogy to me. While another student said that when she first came to the academy and saw the older students' work: "All I could think was, how do you make things like that?'"

Interestingly enough, the academy's present-day curriculum has changed in several important ways since the Antwerp Six studied in Antwerp. As one instructor explained to me, the majority of these changes were initiated in the 1980s when the teaching staff began encouraging students to create highly personalised design collections that challenged popular assumptions about what counts and what does not count as fashion. This included the introduction of several new curricular activities that centred on the development of design concepts. These exercises were not meant to merely teach students how to interpret contemporary fashion trends. Instead they were designed to teach students how to develop research techniques for developing designs that could potentially inspire future design trends and fashion movements (Nicewonger 2016). As a result of these practices, present-day student collections *often* employ visually compelling and experimental forms as well as the use of decontructivist design techniques and exaggerated volumes.[8] This is a point that fashion historian Sarah Heynssens further connects back to the school's pedagogical philosophy when she writes:

> The emphasis is on the artistic design process and less on the technical development of the craft. This study program is perceived as a creative cocoon in which students can enjoy the freedom they need to delve deeply and uninhibited into the design process.
>
> (2013: 25)[9]

This curriculum is organised into four years of instruction. The first three years are part of the baccalaureate program, while the fourth year is considered a master's level course. However, the curriculum is structured in such a way that each year builds upon the former.[10] There are, in other words, techniques and methods introduced in the first year that students will further expand upon in their second, third and fourth year. Accordingly, students encounter the notion of atmosphere in all four years of instruction. But sometimes these encounters will focus largely on learning how to identify sources of inspiration, such as in the first year when students are introduced to basic research practices. In other contexts, like when students are working on the actual wearable forms of their design collections, the concept might be employed to evaluate whether a hem or colour scheme has the right feel or atmosphere of the particular kind of design collection that they are trying to create.

However, despite atmosphere's prevalence in the academy's day-to-day teaching activities, while I was conducting fieldwork, teachers never formally defined the concept. This was in part because the majority of the school's pedagogy occurs in one-on-one interactions.[11] And since there were no formal lectures or assigned readings, students must learn to employ the concept through curricular activities and interactions with their instructors (Nicewonger 2016).[12] Thus, it is through situated engagements with their teachers, as well as through students' own experiments with various techniques for producing design concepts, that the category of atmosphere gains meaning and value (Lave and Wenger 1991). In other words, the meaning and value of atmosphere is generated by the actions that go into its production.

Writing about the ethnographic study of design processes, Keith Murphy argues that ethnographers need to "locate creativity *between* designers rather *in* them" (Murphy 2015: 27). In making this argument Murphy is drawing attention to how designers as well as the materials, methods, tools, and institutions in which they work in, exert a force on the design process (ibid.). The job of the designer, as Murphy argues, "is not to impose form onto matter, but instead to guide the *becoming* of things by channelling 'fields of force and currents of material' in considered ways that shape and fashion a novel configuration of existence" (Murphy 2015: 47).[13] These "forces" and "currents" are produced through a designer's engagements with the material-social world (ibid.). Thus, in the context of the academy the category of atmosphere helps facilitate students' understanding of the meaning of these processes and their creative value.

I will explore this further in the following sections that outline four modes of practice through which student designers learn to define the atmosphere of their design collections. In doing so I seek to offer a generalisable schematic based on my fieldwork observations. This schematic should not be taken as a recipe or formula for producing atmosphere. Instead it should be read as a descriptive break down of how students learn "to follow the forces and flows of [the] material[s] that bring the work into being" (Ingold 2013b: 96).

Mode one

The first mode revolves around the search for inspirational sources that students will later use to conceptualise their imaginary worlds. During this time the category of atmosphere is not frequently talked about, though sometimes students will reflect back upon the research process by saying that they chose a particular image or object because of its atmosphere.[14] Nevertheless, atmosphere is still intimately connected to the search for inspirational sources, even if that connection is largely implicit, since it is through this practice that students learn to identify and work with the affective and sensorial qualities of the material and visual sources that they will eventually use to create their imaginary worlds, which is in turn how they come to define the atmosphere of their design collections (i.e. Mode 3).

Inspiration can come in many forms. But the most common sources include both found and made objects as well as images and samples of written work, like lyrics, poems and quotes. That said, photos and related imagery are by far the most popular as they are generally more accessible and easier to work with than objects and text.

When they search for inspirational sources, students in the second, third and fourth year begin looking for source materials at the start of the school year. In the first year, students are introduced to basic research principles and so while they are also actively searching for sources of inspiration, their engagements with inspiration are largely dictated by the objectives of curricular activities rather than by the development of a design collection.[15] In either case, the search for inspiration can involve a wide range of activities.

For instance, one of the students I worked with began his search by taking photographs of religious regalia and buildings in his home country. He then conducted extensive research on religious costumes, artefacts and imagery. Additionally, he read articles and news reports on contemporary religious issues so that he could identify further source materials as well as gain a deeper understanding of his topic.

Other students, however, might begin their search in an archive. And still, others might start by searching online for images that reflect a particular "feeling" that they are interested in exploring, which may in turn lead them to visit an outdoor market or to attend an art exhibition and so forth and so forth. What this means, then, is that there is no singular starting point or site for identifying sources of inspiration. Instead, students are socialised to attribute inspiration to *something* they experience through their interactions with the social world. This is why teachers strongly encourage students to collect as diverse a range of source materials as possible. Take for instance the following interaction where an instructor explains to a student why the images he has selected are insufficient:

> A student has laid out 8 to 10 images, which his teacher studies for a few seconds before asking: "Are these all the images you have?" "Yes," replies the student, "I'm still looking for more images". To which the teacher responds: "The more material you have the more it's working in your head. It will stimulate you to imagine".

As this interaction illustrates, at the academy students are taught that design ideas are not born out of thin air, but instead emerge through bodily engagements with the sensorial affordances of source materials and interactions. For many students this model is counter-intuitive because they often associate the generation of ideas with the internalised cognitive abilities of individuals (Ingold and Hallam 2007). Students also find it difficult to get past their preconceived ideas about fashion. A case in point was a classroom interaction I once observed where a student had an assortment of pictures that included an image of a garment by a well-known designer. Upon seeing the picture the design teacher immediately threw the image to the side, while firmly telling the student: "This says too much!"[16]

This image, in other words, was deemed an invalid source of inspiration because it reflected the creative insights of another designer. As a consequence teachers and older students often say that it is better to collect as much as possible without thinking too much about what your collecting: "If you just let yourself go, things will start to emerge and those usually are the best ideas," was how one student put it. Or as another student explained: "[If] I don't have anything in my mind. . . . I can do it more freely and [create] whatever I want".

In summary, the search for inspirational sources requires students to attune to the affective qualities of inspirational sources, not their representational meanings (Stewart 2011, 2006). Thus, the search is not about collecting *things* – pictures, objects, etc. Rather it's about the atmospheric feeling or animate forces that students experience when they come into contact with inspirational sources.

Mode 2

The second mode of practice requires students to further evaluate their source materials. It also provides students with an opportunity to play with the shapes, textures and affordances of the inspirational sources that they have previously collected. These experiments can take on any form, but one of the most common is to edit images and create collages. Take for instance the following interaction between a teacher and a student.

A student stands with her teacher before a series of images that she has laid out on top of large white desktop. While looking at the images, which are from a film, she says to her teacher: "I don't want it to be [to] strongly [referred to]. I wanted to refer to [the film], but otherwise . . . I wanted to make it my own". To which the teacher replies while pointing to one of the images with a person in it: "By taking her [as a source of inspiration] you get something totally different. The atmosphere is totally different. It's different".

In this exchange, the concept of atmosphere is introduced as a conceptual tool for helping students experiment with the affective qualities of their source materials. Editing is also a way for students to add particular parts of one source material to another, which in turn allows them to build something new out of their collected materials. In doing so, they are able to formulate new relationships between objects, colours and imagery. To help students edit and sort through their source materials, teachers often engage in forms of design talk that are meant to help students flesh out the atmospheric qualities of inspirational sources. In doing so, teachers are socialising students into embodying perceptual, tactile and procedural dispositions for generating design forms (Keller and Keller 2003; O'Connor 2007). Take for instance the following interaction where an instructor uses the term "stronger" to comment on the quality of a student's research images, which she is using as inspiration for her design.

A student stands in front a long white table where she has laid out approximately 20 images from varying mediums of all different shapes, sizes and colours.

Upon evaluating the images, the teacher picks up a picture showing a competitive swimmer breaking through the water's surface as he races towards the finish line. Referring to the image the teacher says to the student: "This image is not strong enough". The teacher then begins placing her hands across varying sections of the photo and says: "but you can play with the parts of the picture [that are strong]".

As this teacher's critique illustrates images do not in and of themselves generate inspiration. Instead, creative insights are operationalised through the methodological practices that designers use to select and transform found objects into source materials. Through such practices sources of inspiration become a medium for blending aspects of the material world with a designer's own subjective interests (Duranti and Burnell 2004; Hutchins 1995). As such, the actual subject of the image is less important than the affective responses that it might produce through later analysis. For these reasons, it does not matter whether the initial subject of inspiration is a photo of a fuzzy little white kitten or a painting of Dante's Inferno; as long as the image elicits a sense of excitement, curiosity or introspection, it is considered a valid source of inspiration and thus a strong image.

While classroom interactions helped me understand the social sanctioning of individual selections, they did not illuminate the experiential nature of distinguishing between valid versus invalid sources of inspiration. Thus, in an effort to grasp a deeper understanding of these practices, I attempted this process on my own.[17] This experience is captured in the following excerpt where I share a group of images that I collected with two advanced students, whose evaluations of my work reveal further insights into the cultural logic behind the analysis of inspirational sources.

TODD: I found these pictures of blue things and they were all sitting next to each other and I had this instinct that they might make a good [design concept]. . . . I thought these are interesting because there was this clear relationship between the particular colour of blue and the forms and shapes in each of the pictures. But because the forms and shapes weren't really recognisable or definitive I thought I could maybe use them to create some kind of creationist myth. And these were the ones that didn't work . . . because it was too easy to identify them to a certain place and time.

VICTOR: I think it's a good start . . . all of these images are powerful. . . . When you collect images it's just not about the colour it's about what they mean to you . . . I think these images are super interesting and it's nice how you approached it so individualistically and thought about why you felt really strongly about each of them [as] you put them together.

TODD: So I started with an idea, is that okay?

LUNA: That's just how you are. For me for example I just collect pictures all the time and . . . then I go through them and find some direction.

VICTOR: Yeah there's no one way of doing it.

I draw on this exchange to illustrate how students are socialised to construct associations between affect and imagery, that in turn help them identify methods for producing meaningful shapes, construction techniques and aesthetic patterns. As my experience suggests there is no definitive formula or schema for describing the distinction between strong versus weak images. A point Victor emphasises in our interaction when he says: "there's no one-way of doing it". But while these practices lack a clear methodology, in specific contexts – like the one created through my interaction with Victor and Luna – designers are able to jointly talk about and analyse another person's selections because the category of atmosphere is methodologically tied to a broad set of principles for conducting research, which to this point I have outlined two general practices (Murphy 2012). The following section explores a third mode where students turn their research materials into imaginary worlds.

Mode 3

Figures 9.1–9.3 includes three images from a second-year student's research portfolio. When viewed in this way these images resemble what is popularly known in the fashion industry as a mood board. As fashion scholar Tracy Cassidy writes: "Mood boards provide a 'space' to arrange the collected visuals in a meaningful manner to the designer that enables the flow of thoughts, inspirations, and creativity for design outcomes – products" (2011: 230). In this way mood boards become conceptual resources that designers can consult through the design process in

Figure 9.1 Research portfolio content

Figure 9.2 Research portfolio content

Figure 9.3 Research portfolio content

order to address varying design problems or technical issues (ibid.). Subsequently, the creation of mood boards is widely practiced in the fashion industry. But interestingly enough, while I was conducting research at the academy I rarely, if ever, heard the term being used. Instead concepts like "atmosphere" or references to one's "research materials" and "imagery or design world" were evoked in day-to-day conversations as well as in informal discussions outside of the classroom. Consequently, in this section I want to describe some of the key principles that students draw on while creating their design worlds.

At the academy, world-making is a design process that involves teaching students how to craft highly illustrative social scenes using the found objects, images and iconographic forms that are described in modes 1 and 2. When these inspirational sources are brought together they are referred to as the imaginary world of a particular design. It is in these elaborately illustrated scenes that the imagined subjects of a design come to life. And while the materiality of these imaginary spaces are important, it is the atmospheric feeling of these worlds that designers are most interested in, because it is through the social constitution or the making of these worlds that students come to understand and define the atmosphere of their collections. The processes underlining the creation of these worlds, however, are highly circuitous and for that reason students often say that they never know what they are doing until they've actually finished their designs.

The unfinished-ness of the design process, in fact, is key. Consider a conversation that I had with a third year design student named Filipe about the atmosphere of his collection. Inspired by an image he glimpsed while attending a concert of a father carrying a child on his shoulders, he translated this experience not as a direct engagement with fatherhood, but in terms of a general interest in the aesthetic principles of sculpture and food. As he explained:

> [When I saw it] I thought it was an interesting silhouette because [of its] volume . . . I [then began] thinking [about how] I am not going to have children because biologically I will not be with a woman. . . . This is when I started to think [in the abstract about] how I could represent the man with the child without actually showing the man with the child. [As] I started to draw I [began to] understand that the point of the man having a child is a connection with the roots. That's how I saw that the two things were the same [and that] I could actually [achieve this] same feeling if I [worked] with primitive sculpture and food.

As Filipe's explanation illustrates the creation of a world through which his design could come into being was not about producing a linear concept of knowledge or reason. Rather it was about embodying the moral imaginaries of the subjects inhabiting his design world (Nicewonger 2011). Imaginary worlds, thus, play an important role in teaching students how to generate sources of inspiration that they believe will give birth to new forms, patterns, and aesthetic compositions. It is through this activity that students develop a perceptual sensibility that will guide them as they begin developing the actual designs of their collections. In this way the temporal processes underlining the creation of these worlds teaches students how to enter into and experience another time and place, a locale where the material and aesthetic flows of imagined people, creatures, and even cosmologies co-exist and interact with one another.

The technique is also one of the ways in which the personal view of the designer is given space to grow and develop. But developing a personal viewpoint does not mean that intersubjectivity is not important (Murphy 2012). In the context of the academy, individuality is a means for producing a subject position that is open to following a hunch or inclination to fruition, even when the hunch seems impossible to realise. In the process of working with these imaginary worlds, students come to embody certain sensibilities that shape their sensorial interactions with the tools of their trade – the sketching paper, the pens, the water colours that they draw their first design sketches on (Ingold 2013b). As one student explained:

> A lot of things happen unconsciously, you pick a subject, [and] you're out in the world, in an atmosphere. You have all your pictures. You have all your information. Then you start designing. . . . [But] it's not that you try, it's just you're drawing and at that moment I think your brain, is kind of – yeah I don't know how to say it – you draw naturally and then in the end, it's you. It's you that is coming out of you.

As this student's quote makes clear, the translation of ideas into design forms is never literal. Instead, value is produced through sensual engagements with both materials and the spiritual domains of social life: "I just loved it;" "I used it because I felt it. . . . " or "It feels wrong". "I feel like it's going to turn into something." In such ways the affective qualities of aesthetics are given social significance not just by the designer, but by the properties of the textiles, shapes and colours used in the design. Through these interactions, a designer is able to gauge the potential exchangeability of their designs, as they begin to take form and make their way into the world.

Mode 4

This last mode is where the students begin to work on the actual wearable designs of their collections. This process involves first sketching out design ideas, which students discuss with their instructors and revise, usually several times, before they begin their first attempts at creating a toile (i.e. mock up of the design). Subsequently, during this process students and teachers often compare and contrast their design sketches to the toiles they are working on. If there are discrepancies or other details that need to be elaborated upon and/or amended students will turn to their research portfolios and look for further inspiration. In this way atmosphere helps students attune their design practices to the conceptual work they developed during modes 1–3.

This is important because as one teacher explained, once students begin creating the actual material versions of their designs, it's easy for them to lose sight or to become confused about how the conceptual research that went into their design's initial development figures into the actual materialisation of that work. In part, this is because students are not necessarily copying their design sketches when they begin to create their technical patterns or work on their toils, since design sketches do not come with instructions. Rather students have to innovate methods and engineer techniques on their own. But even then, atmosphere is important because it provides students with information that they can use to improvise and develop construction techniques.[18] For these reasons when teachers evaluate students' toiles they consistently ask to see their research materials and design sketches, which they study and compare to the material work being done on the toiles.

This process of course is challenging, which is in part why students in the first three years of instruction are required to complete various curricular activities to help them gain a deeper understanding of the design process. For example, second-year students are asked to make a historical costume at the start of the fall semester. As with the creation of a design collection, this assignment requires second-year students to conduct research and then create a design of their choosing. However, as one student explained, this assignment is not about teaching students how to "re-create garments as they existed in ideal form some 100 or 200 years ago". Rather, it's about teaching students how to "create atmosphere".

This point is further illustrated in the following interaction between a second year student and his design teacher, who is evaluating a toile based on a historical image:

> Mrs Leer the second year instructor stands in front of a model looking at an image of a nun while she rubs the material of the habit through her fingers. "Giovanni, why did you pick this white [for the habit]? It looks like the same white in the picture". Pointing to the image Giovanni replies, "I wanted to make it accurate". Looking back and forth between the picture and the toile Mrs Leer shrugs her shoulders "Yes, but the feeling doesn't make sense, we're not in 17th century. Do you understand?" Giovanni shakes his head in an indeterminate gesture. "No, you're lost," Mrs Leer says as she holds the image between Giovanni and herself. "The colour has to represent the feeling you get from looking at the picture, not what it would look like then. Think about the atmosphere, do you understand?" she asks. "Yeah, I think so. So I should make the habit more dirty?" "Yes, maybe – you have to experiment more, we're making fashion not costumes".

Giovanni's misunderstanding of the assignment provides a useful vantage from which to understand the productive labour of atmosphere in this context. It is through the affective qualities of his colour selections that his design is able to mediate not only interior, individual states but also to generate new alignments of form, function and feeling (Murphy 2012). In these inter-subjective alignments we can begin to see how the concept of atmosphere – in this case expressed through the selection of colour patterns – gives rise to materially novel and yet recognisable forms. These practices, as will be argued in the next section, open up a critical vantage into the ways in which atmosphere can inform not just the analysis of design but also ethnographic experiments in making.

Making, atmosphere and ethnography

Atmosphere, I have argued, is an important pedagogical idiom by which aesthetic innovations are patterned. Through it, design students learn to experiment with the affective qualities of inspirational sources in order to reinterpret material-social relationships through experiments in making. Accordingly, I have sketched out a way to describe the different modes of practice through which students are taught to conceptualise the atmosphere of their design collections. The key point that I want to stress in this final section is that these modes are not only pedagogical forms amenable to ethnographic description. They are also a crucial schema for organising perception and constitute a method for interpreting lived experience (Murphy and Marcus 2013). That is, design methods and ethnographic methods exist in a generative and mutually informative tension.

The first mode of practice, which is reflected in the socialisation of atmosphere, revolves around the search for inspirational sources. What is important about this search is that students identify methods for reflecting on the emergent moods, imaginaries or evolving social events that they encounter while conducting research. The search for inspiration is explicitly not about investigating fashion trends nor about conducting research on consumer patterns. Rather, its primary purpose is to develop methods for catalysing the kind of mysterious embodied force that can result in new trends and fashion movements. Students are taught to explore affective encounters, objectify salient social causalities within them, and then to translate these objectifications into material forms that can be evaluated and graded. Through this loose equation – which of course works both ways – individual "talent" is socially attributed, materially bundled, and re-mystified as sources of inspiration.

The second mode expands on the initial search for inspiration by experimenting and analysing the various moods, sensations and affordances of collected source materials. Through these experiments, students learn to evaluate and relate the efficacy of certain sources to others by looking for patterns that they can later expand upon when they create their imaginary worlds. Subsequently, students often find themselves making nonlinear connections between various sources, which to an outsider might seem extremely subjective or simply unintelligible. But for students – like Isabella, Filipe and Giovanni – the goal is not to represent a social concern or issue but to enact and articulate methods for re-interpreting the causal relationships between aesthetic forms, regimes of value and socially recognisable sentiments. And because these methods are rooted in actions that blend, warp, extend, remove, edit and so forth, they inspire further improvisational acts, including the creation of imaginary worlds (Ingold and Hallam 2007).

In these "worlds", the highly illustrated and richly animated scenes created by students are not meant to depict actual social life. In contrast, their function is to provoke the feelings, sensations and thus atmospheric "forces and flows" of a fantastic ideal (Ingold 2013: 96). That is, design students are expected and taught to repeatedly engage in practices akin to what Kathleen Stewart calls "worlding" (2007: 66, 2011: 445, 2013). In the process of making these worlds, students acquire particular kinds of outlooks and sensibilities that in turn shape how they interact with their design work. These created worlds, in other words, justify and demand novel speculative experiments with material form. If the sensuous creation of these worlds and their atmospheric affordances is a third mode, then a fourth is the public expression and evaluation of these creations by peers and instructors.

The cumulative effect of these modes blurs the lines between the rational and the non-rational. The socialisation of design aesthetics can only be understood in relation to how affect – not epistemology – is the key connective tissue (Stewart 2007). These practices and processes, as I learned through my own attempts at making atmosphere, verge on forms of speculative play. That is, they do more than just produce shapes and patterns. They also provide ways of following hunches, working with unformed ideas, and fragmented observations. They demand that you speculate and thus reflexively analyse events and interactions that already defy coherent description.

This may well be the singular contribution that design may offer to ethnographic methods. Take for instance George Marcus' argument that "classic ethnographic textual form, even as amended since the 1980s and given its learned pleasures, is a very partial and increasingly inadequate means of composing the movements and contests of fieldwork" (2013a: 202). He goes on to argue that there are a number of alternative sites for reimagining ethnographic research, including in art and design communities. These sites are thick, as he contends, with non-textual techniques for studying sociality in ways that build on the central tenets of the anthropological imaginary, even when said practices operate in a radically different way than the "tropes" that often get associated with doing fieldwork (Marcus 2013b: 2, 2000). What happens, he asks, if we experiment with not just writing techniques but with the interventionist and modelling practices of designers and artists?[19]

> Experiment today is thus less about writing strategies and more about creating forms that concentrate and make accessible the intermediate, sometimes staged, sometimes serendipitous occasions of distinctively anthropological thinking and concept work, analogous to prototyping in design and engineering disciplines, in their studios and labs, that occur within the recursive circuits of multi-sited fieldwork still guided by images of moving through natural settings of social action.
>
> (Marcus 2013b: 2)

Brent Luvaas makes a similar point when he suggests that "we should be asking what anthropologists have to learn *with* or *from* design. . . . " (2016)? Design, as he explains, "is about imagining a better world and developing novel ways to help to bring it about" (ibid.). These processes, as this chapter has attempted to highlight, are not just about the creation of things, they involve multiple methods for complex social issues and processes (ibid.). It is these kinds of techniques and the insights they engender – not the finished work of artists and designers – that Marcus contends can compliment and extend the qualitative research currently being carried out in the social sciences by producing new interactional contexts for exploring complex social problems (2013b; Murphy and Marcus 2013; Cantarella et al. 2015). These practices are not meant to replace fieldwork, but rather to open an additional space for experimenting with insights that have been collected using more traditional fieldwork methods (Rabinow et al. 2008).

Likewise, I've argued that we can think about the intentional creation of atmosphere as being not always – or even primarily – about fashion. Like ethnographers, the design students I worked with are not just depicting worlds but inventing and intervening in them, a practice that the notion and technique of producing atmosphere at the academy makes recognisable and thus open to further experimentation.

Notes

1 The fieldwork for this project was supported by research grants from the Fulbright Program and Wenner-Gren Foundation for Anthropological Research (Grant # 7638). The writing of this chapter was in part also supported by a LETStudio Postdoctoral Research Fellowship at the University of Gothenburg. Finally, all the names of the actors mentioned in this essay have been described using pseudonyms and or generic descriptors.

2 At the academy these portfolios were also referred to as: Tendency Books, one's research, a research book, design worlds, and/or a dream world. Because of this diversity, in this chapter I will refer to students' research materials as a research portfolio or design portfolio. I will also describe the social scenes inside these research portfolios as imaginary worlds or design worlds.

3 This conversation was just one of multiple interactions I had with Isabella over the course of the school year (also see Nicewonger 2015).

4 At the academy the atmosphere of a design is also considered in relation to other design elements, such as the balance and volume of a toile's design, see Nicewonger 2016.

5 Martin Margiela is also associated with the Antwerp Six, see Van Godtsenhoven 2013 and Granata 2012.

6 During my fieldwork the student body at the academy included students from all over Europe as well as from various countries in Asia, North and South America, and Australia.

7 The subject of avant-garde is a contested category. Nevertheless, students often used this term to describe the school's curriculum. For an analysis of how this label figured into the contemporary history of Antwerp-fashion see Martinez 2007 and Teunissen 2011.

8 Deconstructed fashion refers to a design movement in fashion. As fashion studies scholar Francesca Granata explains: Belgians designers associated with this movement "play with the function of clothes and accessories, their unfinished nature and denial of seamlessness" (2012: 4).

9 For further information about the methods and philosophy underlining this training see Verschueren 2015 and Nicewonger 2013.

10 For more about the structure of the curriculum see Nicewonger 2013 and 2011.

11 By lecture I am referring to a formal presentation on a subject matter. From time to time teachers will gather students together and provide information about an assignment or school related matter. But the majority of instruction I observed took place in meetings between a particular student and their teacher. The students, however, do take a series of non-design courses at the academy that involve lectures (e.g. history, sociology, and art theory).

12 To some extent students said that they learned about the design process by talking to and observing their classmates.

13 See Ingold 2014, who has also made important contributions to this discussion.

14 Before they started working on their collections students in the third and fourth year often had a sense of the kind of atmosphere they wanted to work with. In part this was due to the fact that these students often spent time over the summer conducting research for the following year. They also often said that after studying at the academy for a couple of years your senses become very alert. As a result it was difficult for them to not collect items, particularly if they happened to wander into an inspiring art exhibition or if they were traveling in a foreign country.

15 In the first year, students are given a series of curricular activities that introduce them to certain aspects of the research process. They do not design a collection but create an experimental dress and skirt, which they exhibit at the end of the year during a school fashion show. In the second year, students are responsible for approximately four designs, while student collections in the third year include eight to nine designs. In the final year students are expected to create a full collection that include a minimum of 12 silhouettes.

16 This interaction was not unique; I witnessed other similar exchanges. In Nicewonger 2013 I describe another interaction, this time involving a jacket, which produced a very similar kind of response from the design teacher.

17 In Nicewonger 2015 I reflect on my personal experience searching for inspirational sources in an experimental essay that explores the world making practices of fashion designers and livestock producers.

18 As students conduct research and create their imagined worlds it is not uncommon for them to identify design techniques that they did not previously know about, such as ways of altering color patterns; methods for tailoring a jacket; or techniques for creating varying prototypes.

19 Recently there has been a burgeoning body of work exploring the interrelationship between design and anthropology. This includes edited volumes reflecting on the anthropological investigations of design (Clarke 2010; Gunn and Donovan 2012; Otto and Smith 2013). There is also a number of important works theorising design's relationship to anthropology (Rabinow et al. 2008; Suchman 2011; Murphy and Marcus 2013; Marcus 2013; Cantrella et al. 2015). See also related writings by Guggenheim et al. 2013 and Salter 2015.

Bibliography

Cantrella, L., Hegel, C. and Marcus, G. E. 2015. A Week in Pasadena: Collaborations Toward a Design Modality for Ethnographic Research. *Field A Journal of Socially Engaged Criticism* [Website]. Available at http://field-journal.com/issue-1/cantarella-hegel-marcus

Cassidy, T. 2011. The Mood Board Process Modeled and Understood as a Qualitative Design Research Tool. *Fashion Practice* 3(2), 225–252.

Chumley, L. H. 2013. Evaluation Regimes and the Qualia of Quality. *Anthropological* Theory 13(1/2), 169–183.

Clark, H. 2012. Conceptual Fashion. In *Fashion and Art*, edited by A. Geczg and V. Karaminas. Berg: London, 67–76.

Clarke, A. (ed.) 2010. *Design Anthropology: Object Culture in the 21st Century*. New York: Springer.

Duranti, A. and Burrell, K. 2004. Jazz Improvisation: A Search for Hidden Harmonies and a Unique Self. *Richerche di Psicologia* 3, 71–101. Available at www.sscnet.ucla.edu/anthro/faculty/duranti/publish.htm [Accessed 11 December 2010].

Granata, F. 2012. Deconstruction Fashion: Carnival and the Grotesque. *Journal of Design* History 26(2), 182–198. Available at http://jdh.oxfordjournals.org [Accessed 28 August 2014].

Guggenheim, M., Kraeftner, B. and Kroell, J. 2013. 'I Don't Know Whether I Need a Further Level of Disaster': Shifting Media of Sociology in the Sandbox. *Distinktion: Scandinavian Journal of Social Theory* 14(3), 284–304. Available at www.tandfonline.com/doi/abs/10.1080/1600910X.2013.838977 [Accessed 4 February 2015].

Gunn, W. and Donovan, J. 2012. Designing Anthropology: An Introduction. In *Design* and *Anthropology*, edited by W. Gunn and J. Donovan. London: Ashgate, 1–14.

Haraway, D. 2011. Speculative Fabulations for Technoculture's Generations: Taking Care of Unexpected Country. *Australian Humanities Review* 50. Available at www.australianhumanitiesreview.org/archive/Issue-May-2011/haraway.html#bio [Accessed 22 January 2017]

Hayward, E. 2012. Fingeryeyes: Impressions of Cup Corals. *Cultural Anthropology* 25(4), 577–599.

Heynssens, S. 2013. Between Avant-Garde & Tradition. In *Fashion Antwerp Academy* 50, edited by B. De Man, A. Haegeman and J. Van Doorselaer. Lannoo: Belgium, 13–28.

Hutchins, E. 1995. *Cognition in the Wild*. Cambridge, MA: MIT Press.

Ingold, T. 2013a. The Materials of Life. In *Making: Anthropology, Archaeology, Art and* Architecture. London and New York: Routledge, 17–31.

Ingold, T. 2013b. Bodies on the Run. In *Making: Anthropology, Archaeology, Art and* Architecture. London and New York: Routledge, 91–108.

Ingold, T. 2014. The Creativity of Undergoing. *Pragmatics & Cognition* 22(1), 124–139.

Ingold, T. and Hallam, E. 2007. Creativity and Cultural Improvisation: An Introduction. In *Creativity and Cultural Improvisation*, edited by E. Hallam and T. Ingold. Oxford: Berg Publishers, 1–24.

Keller, C. and Keller, J. 2003. Thinking and Acting with Iron. In *Understanding* Practice*:* *Perspectives on Activity and Context*, edited by S. Chaiklin and J. Lave. Cambridge: Cambridge University Press, 125–143.

King, K. 2011. *Networked Reenactments: Stories Transdisciplinary Knowledges Tell*. Durham, NC: Duke University Press.

Lave, J. and Wenger, E. 1991. *Situated Learning: Legitimate Peripheral Participation*. Cambridge: Cambridge University Press.

Luvaas, B. 2016. Design and Temporality: Deviation. Correspondences. *Cultural* Anthropology, Website. Available at www.culanth.org/fieldsights/893-design-and-temporality-deviation [Accessed 20 June 2016].

Marcus, G. E. 2000. Introduction. In *Para-Sites: A Case Book Against Cynical Reasoning*, Late *Editions 7*, edited by G. E. Marcus. Chicago: University of Chicago Press, 1–13.

Marcus, G. E. 2013a. Experimental Forms for the Expression of Norms in the Ethnography of the Contemporary. *HAU: Journal of Ethnographic Theory* 3(2), 197–217.

Marcus, G. E. 2013b. Prototyping and Contemporary Anthropological Experiments With Ethnographic Method. *Journal of Cultural Economy*. Available at http://dx.doi.org/10.1080/17530350.2013.858061 [Accessed 12 March 2014].

Martinez, J. G. 2007. Selling Avant-Garde: How Antwerp Became a Fashion Capital (1990–2002). *Urban Studies* 44(12), 2449–2462. Available at sagepub.com [Accessed 18 August 2014].

Menkes, S. 2013. A Rare Reunion for the 'Antwerp Six'. *The New York Times*, 17 June. Available at www.nytimes.com/2013/06/18/fashion/a-rare- reunion-for-the-antwerp-six.html?_r=0 [Accessed 4 September 2014].

Murphy, K. M. 2012. Transmodality and Temporality in Design Interactions. *Journal of* Pragmatics 44, 1966–1981.

Murphy, K. M. 2015. *Swedish Design: An Ethnography*. Cornell, NY: Cornell University Press.

Murphy, K. M. and Marcus, G. E. 2013. Epilogue: Ethnography and Design, Ethnography in Design . . . Ethnography by Design. In *Design Anthropology: Theory and Practice*, edited by W. Gunn, T. Otto and R. C. Smith. New York: Bloomsbury, 251–269.

Nicewonger, T. E. 2011. *Fashioning a Moral Aesthetic: An Ethnographic Study of the Socialization of Antwerp Trained Fashion Designers*. PhD dissertation, Columbia University.

Nicewonger, T. E. 2013. The Fashion Designer as Artist: Reflections on 50 Years of Pedagogy. In *Fashion Antwerp Academy 50*, edited by B. De Man, A. Haegeman and J. Van Doorselaer. Lannoo: Belgium, 171–195.

Nicewonger, T. E. 2015. Boundary Objects Revisited: A Comparative Analysis of World Making. In *Avant-Garde Fashion Design and Livestock Husbandry. Mind, Culture, and Activity* 22(2), 152–167. Available at http://dx.doi.org/10.1080/10749039.2015.1026911 [Accessed 12 June 2015].

Nicewonger, T. E. 2016. Exploring Creativity: An Ethnographic Approach to Studying Fashion Design Pedagogy. In *Fashion Studies, Research Methods, Sites and Practices*, edited by H. Jenss. London and New York: Bloomsbury, 117–133.

O'Connor, E. 2007. Embodied Knowledge in Glassblowing: The Experience of Meaning and the Struggle Towards Proficiency. *Sociological Review* 55(1), 126–144.

Otto, T. and Smith, R. C. 2013. Design Anthropology: A Distinct Style of Knowing. In Design *Anthropology: Theory and Practice*, edited by W. Gunn, T. Otto and R. C. Smith. New York: Bloomsbury, 1–32.

Rabinow, P., Marcus, G. E., Faubion, J. and Rees, T. 2008. *Design for an Anthropology* of *the Contemporary*. Durham, NC: Duke University Press.

Rutherford, D. 2016. Affect Theory and the Empirical. *Annual Review of Anthropology* 45, 285–300.

Salter, C. 2015. *Alien Agency: Experimental Encounters With Art in the Making*. Cambridge, MA: MIT Press.

Stewart, K. 2006. Cultural Poesis: The Generativity of Emergent Things. In *Qualitative* Research, *Third Education*, edited by N. K. Denzin and Y. S. Lincoln. Thousand Oaks, CA: Sage Publications, Inc., 1027–1042.

Stewart, K. 2007. *Ordinary Affects*. Durham, NC: Duke University Press.

Stewart, K. 2011. Atmospheric Attunements. *Environment and Planning D: Society and* Space 29, 445–453.

Stewart, K. 2013. Studying Unformed Objects: The Provocation of a Compositional Mode. *Cultural Anthropology*. Available at https://culanth.org/fieldsights/350-studying-unformed-objects-the-provocation-of-a-compositional-mode [Accessed 5 July 2016].

Suchman, L. 2011. Anthropological Relocations and the Limits of Design. *Annual* Review *of Anthropology* 40, 1–18. Available at www.annualreviews.org [Accessed 3 July 2012].

Teunissen, J. 2011. Deconstructing Belgian and Dutch Fashion Dreams: From Global Trends to Local Crafts. *Fashion Theory* 15(29), 157–176.

Tsing, A. L. 2015. *Mushrooms at the End of the World: On the Possibility of Life in* Capitalist *Ruins*. Princeton, NJ: Princeton University Press.

Van Godtsenhoven, K. 2013. The Wonder Years of the Antwerp 6 + 1. In *Fashion* Antwerp *Academy 50*, edited by B. De Man, A. Haegeman and J. Van Doorselaer. Lannoo: Belgium, 65–105.

Verschueren, A. 2015. *Doing Fashion Design: Autoethnographic Reflections in Dialogue With Architecture*. Doctorate of Arts dissertation, University Antwerp.

Williams, R. 1977. *Structures of Feeling, in Marxism and Literature*. Oxford: Oxford University Press, 128–135.

10 Living atmospheres

Air, breath, song and mutual constitution in experimental theatre

Caroline Gatt

Introduction

How are we moved by the atmospheres of particular performances? Are we moved because the atmospheres we dwell in not only touch us, but permeate us as well? Can an actor deliberately generate an atmosphere? Should we consider an atmosphere a medium? How can we know an atmosphere? In this chapter, I pose these questions in relation to the craft of performance through the ethnographic lens of working with experimental theatre makers.

Recently, Throop has associated the notion of atmospheres with permeation (2014: 70), a totalising mood that 'covers over' perception (ibid 2015: 57). In this thinking, our bodies are permeated by the atmospheres we move through and dwell. The forces, or constituents that make up the airs, waters, soils, muds and other substances we move through and live in can be dispersed and appear totalising and are thus considered atmospheric. Herein I discuss the process whereby air, and different airs in different places and situations, affect the person-organism because these airs touch us and have different compositions. These airs and the atmospheres they participate in creating can be understood as constitutive, as participating in generating life.

Crucially, for the notion of atmosphere to be fruitful certain, contradictions within the broader theories these discussions are part of need to be addressed. First, the ethnographic work with song and experimental theatre offers a critique of the implicit notion of a universalised human body, embedded in the phenomenological overtones in discussions on atmosphere in anthropology. Second, since the notion of atmosphere is used to highlight how human experience is immersed in the world, I will argue that the insistence on considering an atmosphere as *a medium* contradicts the ecological thrust that the very notion of atmosphere puts into relief. Simply put, much writing about atmospheres actually reinforces a mind-body dichotomy. The concept of person-organism (Ingold 2000b) acts as a heuristic to identify when a contradiction between a stated ecological or immersive approach and an implicit mind-body and body-world dichotomy is present.[1] The person-organism is a term that calls anthropological attention to the human as a whole being, simultaneously including the social aspects of human life and experience and also their physiological constitution, development, ailment, healing and biological history within one sphere of enquiry, rather than split into the separate fields of socio/cultural and biological anthropology. Third, I argue that to know an atmosphere implies being able to participate in making it and characterising atmospheres as a *pre*conditional medium precludes us from asking questions about how they can be manipulated. I ground this argument in an ethnographic reflection on the actor's work to show how not only do humans emanate into and absorb a particular atmosphere, but that by emanating we can begin to shape aspects of our atmospheres deliberately. Shaping an atmosphere is central to the actor's work and as such offers insight into how we know atmospheres. In this chapter, my ethnographic treatise of the actor's work on vibrational song and Taijiquan suggests that attentional practices grow different knowing bodies. Therefore how we set out to know atmospheres in the first place contributes both to the conceptual *and* to the physical perceptual apparatuses available for one's enquiry. It is for this reason that theory is inseparable from ethnography, for what is included in ethnographic descriptions in the first place also depends in part on what 'theories' participated (with many other things) in developing the researcher's attention.

For this chapter I focus on the work of two theatre makers, Jerzy Grotowski and Ang Gey Pin, as well as on my own training and investigations. The form of theatre I work with is considered post-theatrical, experimental theatre, research or laboratory theatre (Magnat 2014; Spatz 2015; Gatt 2011; Camilleri 2013). Ang Gey Pin is a theatre maker, who has performed, directed and taught in Europe, North and South America and Asia. She is considered one of the theatre makers developing Grotowksi's legacy due to her long-term involvement with the Jerzy Grotowski and Thomas Richards Workcentre (Magnat 2014; Tatinge Nascimento 2010). I consider Ang's work through our collaboration for my current research as part of the project *Knowing from the Inside*,[2] through my participation in her Sourcing Within Worksessions,[3] as well as through her writings and theatre scholars' writings about her work.

Atmospheres and the actor's presence

Many of the workshops that I have participated in focused on the actor's presence. This presence is understood as tangible. One director gave the example of a person first entering a room full of people, at a party or in someone's home. The atmosphere in the room is palpable. Almost fundamental to the actor's work is the cultivation of their presence in a way that reaches out beyond the actor's self to meet others. How does an actor's presence reach out, meet others or enable encounter? Key figures in 20th-century theatre have looked to breath to explore this question. Antonin Artuad came to think about theatre as the "frightful transfer of energy from body to body", the vehicle of which is breath (Artaud cited in Nair 2007: 39). Konstantin Stanislawski, well known as the first and still the most influential figure to have developed and articulated specific techniques for acting (Nair 2007; Grotowski 1968, 52014; Spatz 2015) suggested cultivating breathing (Nair 2007: 143). He associated breathing with the emanation of energy rays (ibid). Breathing, for Stanislawski was a means for the actor to send energy inwards and outwards (ibid). Due to soviet censorship, people interested in Stanislawski's work only discovered later that these techniques were based on notions of breath as *prana* in Sanskrit philosophy (Merlin 2015: viii). In Sanskrit philosophy, breath is not an independent element; it is correlated to space and time, part of the very fabric of reality (Nair 2007: 64). There are two kinds of space in this philosophy: universal outer space and the internal space of consciousness. Breath is the vehicle for cosmic awareness, animating the world, including persons. The cosmos is imbued with consciousness and absorbed by the body through breath (ibid).

While Artaud (2010 [1964]) and Stanislawski focus in on breath, Grotowski does not advocate a particular training for breath (ibid: 140). Rather Grotowski explored 'action', because he understood voice to be an extension of the reaction and impulses of the whole person (Flazsen quoted in Magnat 2014: 153). 'Action' is neither a physical nor a psychological state; it is precisely a transcendence of these divisions (Richards 1995). Grotowski lamented how he had been dogged by the association of his work with 'the body' (cited in Magnat 2014: 121).

> [I]n his form of training the body became a conduit, and that consequently it was not the body in and of itself, or, as he put it, the meat (*la viande*) that was important, but rather the flux of living impulses within the body. Hence the point of training was not to foreground the physicality of the body but to render this flux visible.
>
> (ibid)

Grotowski's approach to theatre and to actor training has more to do with what Magnat (ibid:125) refers to as body-in-life. Action understood as an impulse belonging to a body-in-life involves intention, breath, memory, muscles, hope, attention, relations and many other aspects of human action and experience in a way in which it is impossible to separate out breath or air as primary in some way. Doing so would have the same effect, to follow Grotowski's reasoning, as foregrounding the body for itself rather than its participation in action. In fact Grotowski (1968: 32) characterised his approach as 'poor theatre'; one that strips away all the paraphernalia of theatre: lighting, costumes, recordings, effects etc. . . . and is concerned squarely with the craft of the actor in *encounter*. In some forms of theatre, even in Grotowski's earlier phases of work, the encounter is between actor and spectator (ibid). In his later work and in experimental theatre more broadly, the actor cultivates the capacity to encounter in a much broader sense.

Ang calls the shifting groupings of actors that she created 'Sourcing Within'. The name 'Souring Within' refers to the search for encounters with one's personal living impulses, but also with one's heritage, one's ancestors, one's memories. Here encounter, while including meeting both audience and fellow performers, also includes being open to encounters occurring within the self. "*Si può essere coro anche se si è uno, ognuno di noi possiede mille voci*"[4] (Ermanna Montanari 2006: 16, cited in Lembo 2016). Finally, such encounters, whether with internal voices and ancestors, or with fellow performers or spectators, are also partnered with an encounter with the voices of a particular place that the actor is working in. This might seem subtle since we may not be used to considering walls and floors or volumes of air as partners with whom we have encounters. However, in the work of vibrational song the different characteristics of each place become partners in the same way that the singer's different body zones and memories, are partners in making the song and therefore generating encounter.[5] Since the partnerships are so pervasive, everything is a partner or a potential partner, the encounters sought in experimental theatre can be explored as atmospheric.

Breath, wind and *chi*

Having participated in a number of Ang's Sourcing Within work sessions I was initially surprised that although Ang is widely known for her work on song she dedicates much of the session to working with a Taijiquan sequence.[6] Often Ang begins the working day with little games. At the start of each day there is a moment when the participants seem to be almost ready from their various preparations: changing their clothes, collecting their things into tidy piles at the very edge of the room, having one last sip of water, stretching or warming up in ways they have learnt in their own individual training paths. Ang begins moving very quietly in the room without explaining first. Often some participants don't even notice that she has begun. She might move close to someone stretching on the ground and lean in, mischievously, only to draw back as if pulling a light rope briskly between her fingers. She may gently weave a path through the room, making no noise with her feet, as if gathering a wind. She might then stop suddenly but smoothly and squat low on her heels raising her hands up, as if lifting an invisible feather that insists on tumbling from one hand to another as she lifts it higher and higher, only to squat down again effortlessly to raise a different imaginary feather. This is one of her ways of beginning work on the mode of Taijiquan.

In our conversations, Ang has talked about her work with Taijiquan as part of the specific trajectory her artistic work has taken her on, exploring her cultural heritage (also see Magnat 2014; Tatinge Nascimento 2010; Ang 2017). This has led her to developing an approach to song which in her words is "the mode of Taijiquan in vocal work too" (cited in Gatt 2015: 340). In trying to understand what Ang might have meant when she wrote this to me, I began recalling the work sessions and looking through my field notes. What is *the mode* of Taijiquan? How is it or could it be present in vocal work? At first I couldn't separate Taijiquan from the work of learning the actual sequence, although my instincts were telling me that clearly when Ang for instance playfully begins the work sessions, she is also exploring the 'mode of Taijiquan'. Then I remembered how during both our individual work on our solos as well as during some group improvisations, Ang had often said that I tend to 'cut'. This had baffled me. So as part of my individual research in my studio in Aberdeen, I dedicated my own training to working out how not to 'cut'. Through this studio work I was led back to what the 'mode' of Taijuquan might be, and I try to explain what I came to understand as cutting, or not cutting, to be in the following passage.

Working on the Taijiquan sequence on my own in the studio in Aberdeen, I recalled Ang, while carrying out a particular move in the sequence, would often say 'arriving, arriving'. Piecing together these moments while I worked on trying not to 'cut', I came to imagine the mode of Taijiquan to be related to the circulation of *Chi* that Ang also refers to while working on the sequence. An action, like playfully pulling an imaginary rope, or following a series of actions in a Taijiquan sequence or even singing a song, would have a particular circulation of energies, or flows. Like a ball on a slow rubber band that travels away from me when I throw it, reaches a moment where the tension of the rubber band is at its extreme and then pulls back. So, although previously I had understood circulation to be like little spirals throughout the body, now I understood circulation to mean *Chi* travelling out of me guided by the shape of the Taijiquan poses. The *Chi* would reach a certain point beyond me, sometimes beyond the walls of the studio, out into the open air beyond, then return to my core only to travel out again along a different path shaped by the next action in the sequence.

In my work, this path of *Chi* is both a visualised and an imagined trajectory, like a line that I can follow with my eyes through the air and the walls. It is imaginary, yet I actually see it with my eyes open even if it is transparent. It is also something I feel in my flesh. When the *Chi* 'arrives, arrives' it is not that I simply imagine it arriving; I *feel* a gentle pull in my torso when it is time to imagine it returning. In all the Worksessions that I participated in Ang never once led a task focused on breath, but on what any participant needed during their work with her. This sometimes included breathing, but just as likely included suggestions to stop singing and to carry on the song in silence.

In other writings about Taijiquan, clear links are made between *Chi* and breath (Hsu 2007, 2009; Kuriyama 1999; Nair 2007). In fact *Chi* is often translated as breath (as in Nair 2007: 58–63). However, *Chi* cannot simply be translated as if breath were a universally transparent process. First *Chi* has been understood as different things in different historical periods in different cosmological, ecological and political environments (Hsu 2009, 2007, see also Yoeli-Tlalim 2010 on *Chi* and *Rlung* winds in Tibetan Medicine). In certain periods, breath was wind that entered the body in different seasons of the year; wind travelling from certain prevailing directions carried specific characteristics affecting the body, some to promote health other causing disease (Hsu 2007). In later periods, coexisting with more politically uncertain moments according to Hsu (2007), the work of medicine was to protect the body from this permeability, and *Chi* became more of an internal wind.

In addition to these historical shifts, in Chinese medicine and the understandings of *Chi* in Taijiquan, the *Chi* circulating in the body is not breath in the sense that Western anatomy understands it: as entering the body through the nose and/or mouth, into the lungs and back out (Kuriyama 1999). Neither is *Chi* oxygen that enters the body through breath that is absorbed into the blood and that then travels to different parts of the body through the circulatory system (ibid). *Chi* travels around the body through a variety of channels which are different to the veins and arteries of Western Medicine (ibid). There is no easy correspondence between breath understood in Western biological understandings and *Chi* found in both Chinese Medicine and the practice of Taijiquan (ibid).

Working in the mode of Taijiquan in the actor's work, including speaking and singing, is understood to develop the body-in-life, and through this ecological training the actor can reach beyond the surfaces of their skin through their *Chi*.

Therefore, *Chi* here is also, but not simplistically, breath.

An aspect of Ang's approach that is directly drawn from Taoist understandings is the path of *wu wei*. In the context of Ang's work, this means that although the work of the actor is to train and cultivate, there is also an important way in which this training is aimed at non-action. Although I have been using the phrase *the actor's work*, and t*he craft of theatre and of anthropology*, these also do not match up easily to understandings of skill. The 'work' of experimental theatre in searching how to channel the songs, air, winds and places we move through and live in highlights how we can be affected and shaped by people and places. However what those atmospheres are, what it is that moves in them and through them is far from being a given. The search for vibrational song in experimental theatre brings attempting to know the specific constitution of different atmospheres into the foreground of attention.

Knowing atmospheres

Experimental theatre, as an experimental craft, explores possible routes to transcend a body-mind dichotomy (Magnat 2014).[7] In a short video about her work with Haitian songs Maud Robart says

> ç'a bouge, ç'a bouge les choses, ç'a bouge les formes des pensees, ç'a bouge les emotions, ç'a bouge quell que chose dans le corps même la nuit après le travail il continue. Le corps continue à danser dans le lit, parceque sont des dances qui reveillent quell que choses comme, s'address quell que chose qu'on en pourrai dire encore instinctuelle, alors ç'a continue, ja, la personne humaine dans sont totalité, e touché, e bougie.[8]

Grotowski worked in great depth with Robart on vibrational song (Wolford 1997: 344). He simply writes that vibrational songs are the types of songs that have persisted over the centuries (Grotowski 1997: 299). The reason for their longevity according to Grotowski is that they carry particular resonances that move the people who sing them (ibid). These songs awaken those who sing them, or those who allow them to be sung through them.

Learning how to listen and sing vibrational song changes the very organism-person of the one learning and singing and therefore how the person comes to know that acoustic atmosphere. Here is an ethnographic example comparable to the actor's work on song, and Ang's work on song and Taijiquan is comparable to the ethnographic example provided by Downey: Through the practice and training of Capoeira, practitioners change the actual physiology of their visual apparatus, among others things (Downey 2007). Where reading for example relies on foveal vision, both Capoeiristas and performers, who need to be aware of the entire performative atmosphere, develop their peripheral vision. Thus "[h]uman physiology and behaviour can be modified, affecting both *what is 'known'* and *what is 'knowable'*" (Downey 2007: 236 italics added).

What I would add to Downey's argument that emerges from working on song and presence in experimental theatre is that the constitution of the atmosphere in which one works also participates in shaping what is known and what is knowable. A similar point is also explored by Shapiro: The chemical atmospheres in prefabricated homes in the US become present not by heightening the senses of those who live there, but by deadening the inhabitants' senses, making persons and animals dizzy, forgetful – their vision blurs (Shapiro 2015). Bodies perceive the chemical presences in the air through the infinitesimally small wounds that the organism struggles to heal when exposed and permeated by them for long periods of time (ibid).

Not only do our practices, such as capoeira and experimental theatre, indeed any daily practice, participate in shaping what we perceive, how we know and what we can know but so do the atmospheres through which we move and in which we live, and they do so in a very tangible way. These points about the cultivation of a person-organism are in fact essential if we are to have discussions that do not drag along assumptions which the very notion of atmosphere in recent anthropological discussions attempts to counter. In fact, we see how this very dualism implicitly informs many approaches to atmospheres, and therefore how we know them.

Materialising atmospheres

The term atmosphere comes from meteorology (Ingold 2015: 73). And yet, the meteorologists have emptied this 'atmosphere' of human experience. Philosophers like Böhme writing on atmospheres in theatre have deprived these atmospheres of air (ibid).[9] In the shift from fluxes to surfaces even air becomes imaginary (Ingold 2007). Irigary (1999) prefigures this argument when she argues that Heideggarian metaphysics has forgotten the air we breathe. We have seen a number of ecological approaches in anthropology that attempt to redress this neglect (see Gatt 2009 for one overview of these). Among these, Ingold has addressed the question of the weather and atmosphere directly. I focus on Ingold's specific argument of the medium in a subsequent section, but here what is important is his critique of the paradigm of embodiment. Drawing on Brenda Farnell's (2000) critique of Csordas (1994) as producing talk about the body rather than from the body, Ingold (2000b) argues that the paradigm of embodiment further entrenches the dualism of body and mind by *residualising* the organism. He suggests that in order to truly develop non-dualist understandings it is the entire organism-person that requires attention.

Working on song with Ang, especially in the context of Grotowski's work with vibrational song, that influenced Ang in many ways, requires the sort of openness to the lived body that Farquhar and Lock (ibid) identify. Both Farquhar and Lock understand the body indeed as an organism-person, one which in addition is not only acknowledged as real, vulnerable and curable, but also as having specific social and historical contingency. To take seriously how song changes atmospheres requires asking both about human experience and the materials that make it possible, without losing sight of the way these bodies-in-life are changed by context, practice and life itself.

A singer needs to be aware that particular rooms will have certain resonances and Grotowski (1968) reminds the actor that achieving vibratory quality in song is not a matter of increasing the volume at which a song is sung. A sound can vibrate in a place even when it is a very quiet one. In fact, in my experience very loud sounds deaden this vibrational quality. Home-Cook (2015) relates how very loud sounds were employed in the performance entitled the *Kursk* at a point in the performance when a character in the play had received the shocking news of the death of his infant son. Deafening sound has the effect of shocking the senses into paralysis, as does blinding light (Maccagno 2016). Initiating particular vibrations in a particular place depends on a careful giving and receiving, in the very process of making the sound which requires a constant adjustment.

Singing close to a wall made of a certain material such as gypsum board makes the stream sound of the song different than if the wall were made of hollow brick, or of impenetrable marble. In my experience, one is not necessarily better than the other; too much reverberation can muddy a sound as much as no reverberation at all can make it sound dead. However, the point is that the singer needs to make small shifts to adapt to these different materials that you are in effect singing *with*. The temperature of the room makes a difference to the quality of sound, but also to the way vocal chords harden or soften; temperature and levels of humidity affect not only breathing but also moisturises the vocal chords differently. With a dry throat and mouth or in a room full of puffy jackets and coats the singer needs to constantly listen to their song as they sing and adapt, and needs tomake small changes to help the song resonate – not only so that it can be heard but so that it can travel and touch listeners in a particular way. In this, who the people are always makes a difference as well.

At the closing event of a workshop in London in 2010, led by Krystian Godlewski, us participants presented a short performance to an invited audience. Amongst the audience was Godlewski's partner and their 3-month-old daughter. In one moment in the performance, we all walked forward in a line, singing a Polish Christmas song taught to us by Godlewski. We learnt it as a rousing song, that according to Godlewski incites people to dancing in his hometown. For a 3-month old, however, this was frightening, and she was restless, uttering little cries. So as we marched hand in hand closer to where the audience sat, mother and baby almost directly in front of me, I felt an urge to soften the song, to protect the baby from the singing belting out from participants from either side of me. At our closest point between line of audience and line of performers, with less than 30cm between us, my spine curved, creating a hollow that mimicked the mothers' seated shaped, and the quality of song became lullaby like. What with the two walls of bodies all around us, those standing, singing loudly and those sitting, the hollow filled with vibrating air between the mothers' body and mine in which the baby sat, took on a different quality, more dream-like. Lullaby-soft song and bodies cocooned me, the mother and the baby, and everything actually became quieter in there. The baby calmed down.

After the performance, the mother, a performer herself, came to speak to me to share once again the moment when we had jointly cared for her baby with body shapes and soft sounds, creating a tangibly different atmosphere like a bubble within the louder atmosphere around us. To do this both myself as the singer, and the mother who was holding the baby responded to each other. As part of the performance, I could not simply stop, bend down abruptly or suddenly sing differently. Sudden changes of the sort would have distracted my fellow performers and possibly disturbed the baby more. The change had to be gradual. In fact, I cannot even identify a single point at which this began or when I became aware of the need to change my action. As we advanced I found myself shifting in relation to the other performers and their song, to the way the song reverberated against the wall behind the audience, to the diminishing distance between the baby, the mother and myself, all the while keeping the song going and constantly listening to whether the changes were having an affect or not. This can be considered a process of attunement.

Ecological attunement

Attunement is a key concept in the current writing in anthropology on atmosphere specifically those in a phenomenological vein. In an article called "Atmospheric attunements", Stewart (2011: 447) specifically employs this notion to convey how people in different situations begin to adapt to changing circumstances, in order not to lose their "sea legs". She describes this as

> a labor of attending to a space opening out of the charged rhythms of an ordinary. There's a pause, a temporal suspension animated by the sense that something is coming into existence. The subject is called to a state of attention that is also an impassivity – a watching and waiting, a living through, an attunement to what might rind up or snap into place.
>
> (Stewart 2011: 446)

Stewart relates attunement to Heidegger's concept of worlding, which is "an intimate, *compositional process* of dwelling in spaces that bears, gestures, gestates, worlds" (ibid: 445 Italics added). For the moment, I just want to emphasise how for Heidegger the process of perception is an essentially generative part of forming the world (Ingold 2000a: 154).

In the case of a person singing, this active aspect of the creation of an atmosphere, of the cusp of becoming, is all the more evident. The singer needs to carry on actively, making the song in order to be able to perceive its effect and to adapt to constantly changing circumstances. If the singer was not singing and listening at the same time, there would be no song to perceive. Singing makes it easier to understand how the process of experience, or in Ingold's terms perception, is a *worlding*. What I mean here is not that the singer *worlds* while the audience receives the world made by the performers. The audience members are co-participants in the generation of particular atmospheres (Home-Cook 2015). Schütz explains this very process of attention to others and the world also as the attunement between musicians (Schütz 1969). The notion of a force field encapsulates this notion of atmosphere, in which there is the capacity to affect and to be affected (Gatt 2011: 30–51; Stewart 2011: 452).

Similarly, Throop (2014, 2015, 2016) also develops a notion of the atmospheric starting from the Heideggerian premise that atmospheres can be understood as moods: as totalising and permeating. As Heidegger observes, "a mood assails us. It comes neither from the 'outside' nor the 'inside,' but arises out of being-in-the-world, as a way of such being" (1927/1962:176, cited in Throop 2014: 69). Adopting this interpretation of the notion of atmosphere, Throop has begun to explore ways to *explain* the links between meteorological atmospheres and the classical phenomenologists' atmosphere as mood. Again, working with the notion of attunement, Throop traces the ever-present mood of despair and hopelessness in Yap, in the Federated States of Micronesia. Here on the one hand the link between mood and atmosphere is already meteorological, "*Ke manigiil yifung ea doba* ("There is excellent weather today") (Throop 2015: 60). Further, he traces how the Yap people's perception of their impending end is closely linked with the actual effects of two devastating hurricanes, one in 2004 and one in 2015. These hurricanes together and the approached termination of a political agreement in 2023 with the US, is all part of the mood of long decline of the island's people since before the Second World War. Throop (ibid: 53) makes this link through Husserl who writes that our attention is pulled or affected by worldly happenings. Throop suggests that a good example of how our attention is pulled by phenomena in the world is the classic test of the incommensurable images of the Rubin Face/Vase (ibid: 52). The Rubin Face/Vase is an image developed by the psychologist Rubin in 1915 (Figure 10.1), in which a dimensional image presents a figure ground reversal experiment. Only one of the images, the faces or the vase, can be maintained at any given moment. That is if you are interested in foveal vision only, I would like to add. Try the peripheral vision that I discussed above and you might have a different result.

In fact, this is an example drawn from conventional psychology, one in which individual and rational understandings of mind and attention are privileged over others.

Here is the main pitfall of phenomenological approaches when it is applied to anthropological projects: phenomenological approaches often assume a universal, individual body (Farquhar and Lock 2007). Lock and Farquhar write about Merleau-Pontean phenomenology, stating that "its focus is on a body presumed to be universal and individual, depicted from the point of view of the subject, embodiment in this tradition can lack both historical depth and sociological content" (Farquhar and Lock 2007: 6). As the ethnographic examples above show, an approach to perception and knowledge based on a universalised 'individual' cannot explain how the actor creates atmosphere through song, among other things. In experimental theatre work with song, first the performers work in a way that brings about collective energies not achievable with an understanding that limits

Figure 10.1 The Rubin Face/Vase image

knowledge to the subjective or the inter-subjective. Theatre scholar Home-Cook (2015) describes the atmospheres generated by performances as sensing selves stretching out. However, he emphasises that it is not only the self of the actor but also the spectator that sounds out. Therefore, even when working alone, the singer comes to work with the walls, the floors, the cavities and shapes of their own body, the air and energies that they direct through breath and action specifically as creative partners.

The universalisation of the body and self in phenomenology (Farquahr and Lock 2007) limits the sorts of questions that can be asked about how persons participate in perceiving and making atmospheres. In fact, although Ingold (2000a, 2000b, 2011) draws on Merleau-Ponty for his theory of perception, his original dwelling perspective is also thoroughly informed by animist ontologies in which personhood is precisely not the biologically universal, 'transparent' individual self of phenomenology. For this reason I have argued that Ingold develops what can be thought of as *ecological* phenomenology (Gatt 2009, 2010, 2011, 2013a, 2013b, 2015). Specifically, the ecological thread of Ingold's thinking leads him to characterise the typically Heideggerian worlding as *emergent mutually constitutive processes between the multifarious aspects of the world.*

Atmosphere as medium

The current work on atmosphere that has most exposure in anthropology has a common thread in its imbrication with phenomenology (Stewart 2011; Ingold 2015; Throop 2014, 2015, See also Zigon 2014). On the one hand, this phenomenological positioning is a constant reminder for anthropologists to pay attention to experience. On the other, a number of assumptions embedded in the philosophical school create certain contradictory positions for anthropology. Here I refer specifically to the notion of atmosphere *as* medium, and further that such a medium is considered a *pre*condition for experience, relations, in short for life.

However, equating atmosphere or air with a medium as precondition precludes asking questions about how these are deliberately shaped, for example by performers singing, and therefore if we agree that mutual constitution is a basis for any perception and knowledge, it also precludes an enquiry about how we know atmospheres. Considering the lives of those living in watery worlds or in soily ones, rather than in aerial atmospheres, helps focus the difficulty of considering the environment, as that which surrounds, as a precondition for relations.

Nigel Rapport's (2003) book on existential power employs Schütz's notions of the lifeworld. Rapport argues that the most readily accessible, if not the only accessible, source of knowledge and perception is the self. One's own individual lifeworld is *transparent*. Other humans' life worlds, in contrast, are the most *opaque* worlds of perception; and the breach from self to other cannot be bridged (Rapport 2003: 218). In the third section of the book Rapport explores how existential power implies that beings have the capacity to transform their surroundings according to will. As an example, he describes the habits of earthworms.

Earthworms, which are best adapted to watery environments, manage to dwell in the relatively dry environment of the soil by transforming it: moisturising it and aerating it with their actions of secretion and burrowing. Thus, in effect, the soil becomes part of the life-supporting system for the earthworm. For Rapport, this integrated system, which the earthworm creates through its actions, entails that the earthworm has extended its self in order to continue living in that environment (Rapport 2003: 221–223)

However, closer examination of Rapport's earthworm example reveals a difficulty. When exploring the relationship of an individual to prior conditions, Rapport (2003: 67) argues, the earthworm *is not determined*, rather their relationship to such external conditions is a dialectical one; An earthworm is in an active relationship with the soil around it and interprets it as a habitable place by transforming it as described above. However, if the earthworm's relation to the atmosphere is an active dialectic, then this environment must also be active. Otherwise the transformative actions proposed by the person or the earthworm would not be dialectical; it would be hierarchical – the medium comes first (as in *pre*condition), and earthworms come second.

The agent – whether human or earthworm – must be seen to exist within a field of *mutually* constitutive relationships in order for efforts exerted to have *any effect* in the world.[10]

In the same way as soil is clearly something that the earthworm not only touches but ingests, the different airs we breath and sing in cannot be thought of as *pre*-existing conditions. The different substances present in air have different effects on our constitution. Therefore in characterising an atmosphere as a medium, understood as a pre-condition for experience and relationships, by definition *pre*-conditions cannot be manipulated or changed by the things or persons composing them. Ingold argues that for instance light is not something we see but what we see in (Ingold 2000a: 263/4/5), and sound is not something we hear but something we hear in (Ingold 2011: 138). Consider if I can change the *pre*-condition in which I exist: either I have to stop existing as I am in order to access this *pre*condition, or I cannot have a relationship *with* it. Clearly neither are correct. Not only does a performer deliberately work to, and sometimes succeeds in changing an atmosphere, but this does not happen by some extraordinary existential occurrence.

Instead of deploying Merleau-Ponty's notion of medium to address the problem, we could easily consider more seriously Ingold's own view of emergence to address the problem. Thus sidestepping the need for the notion of medium or preconditions. Here Ingold's notion of emergence is that the multiple aspects of any given place are continuously in emergence, therefore nothing is ever an object. In this sense, not only do we not have relations with the wind, neither do we have relations between persons, because both wind and persons are not finite objects; all are constantly emerging from their mutually constitutive relations (Ingold 2000a: 162). The shift is from considering there to be things *in* a medium, to considering everything to be part of the mutually constitutive flux (Gatt 2009, 2010, 2013a, 2013b).[11]

The work of trying to sing vibrational song makes this point very clear in the process of singing and sounding. There cannot be a distinction between things and medium when working with vibrational song, because the entire acoustic ecology working as a whole needs to be taken into account. Everything becomes a partner in making that sound (whether in a collaborative mood or a recalcitrant one), not only air, but also the materials of the walls, of the floors, of people's clothes, people themselves as having cavities and densities that resonate differently. Helmreich (2010) reminds us how different human ability to hear is under water, when he considers the notion of immersion in soundscape theory. In the recent discussions I have cited, atmospheres have this immersive quality. And while immersion is a productive way to understand experience – there are no viewpoints from nowhere (Haraway 1991) – the world we are immersed in does not need to be considered as an *untouchable* precondition (pace Ingold).

Conclusion

In the phenomenological vein, attuning to an atmosphere is effectively the perception of worlding, "something . . . *coming into existence*" (Stewart 2011: 446 italics added). Attuning to the atmosphere in this sense is being able to perceive the permeating transformations that are constantly happening in order for the world to come into being.

Here, and elsewhere (Gatt 2011, 2015), I want to suggest that experimental theatre investigation cultivates a way of knowing in which there is no separation between reflexivity and spontaneity. The essence of Grotowski's investigation is one in which "spontaneity and discipline co-exist and mutually reinforce each other" (1968: 121). As an anthropologist bent on taking the investigative work of these theatre makers seriously, I have suggested that this work develops a form of attention which is itself processual, attuned to ongoing emergence. The investigations of experimental theatre develop "reflexive revision in the midst of action" (Gatt 2015: 347). Developing this form of processual attention involves and changes the entire organism-person, as any education of attention would. For instance, learning to imaginatively perceive *Chi* as flowing through one's body, the air in a studio, the walls of the studio to the open air beyond requires and creates a different sort of attention; a form of attention which involves muscular, imaginative, nervous and memorial changes.

Atmospheres might be only vaguely knowable in daily life as Throop comments (2014: 71). However, the work of experimental theatre, specifically through the work on actions and vibrational song, offers a way of knowing dedicated to attending to emergence. It cultivates a way of knowing which is prospective, *in temporal synch with 'worlding'*. This epistemology unsettles our concepts of work, in that what Ang is searching for and what guides the participants in her work sessions to search for is how to be a channel, how to pay attention to the flows that arise within you, that you emit and that you receive. Here the concept of *wu wei* links to atmospheric knowing: To know an atmosphere is to know how one is shaped by an atmosphere. And yet, in order to know how one is shaped takes the work of deliberately trying to shape it, as in the work of the actor extending their presence. This action of non-effort highlights how all the different, emergent, qualities of particular places are partners in the performer's wish to send a song not only beyond themselves but also, when singing very quietly, inwardly, to those very material tissues of bone, cartilage, skin and flesh that vibrate differently with different posture and with different imaginings.

There is a proviso here, as always. To be shaped by an atmosphere is not *automatically* to know it, so although atmospheres might permeate and be part of the mutually constitutive processes that make all of us, we are not automatically aware of these processes. Even if all human knowing is arguably anyway processual, as Ingold has argued "*we know as we go*" (2000a: 229 italics in the original), we have already seen how some atmospheres diminish our ability to perceive their influence on us (Shapiro 2015). And although both Shapiro and I have discussed the small-scale atmospheres in people's homes or in small studio spaces, even intimate atmospheres within one's own body, the notion of atmosphere invites us to think on multiple scales, from these intimate bodily atmospheres to the global atmosphere and anything in between. In fact, Sloterdijk (2009) has written caustically on how atmospheres can be mass weapons, deliberately deployed for destructive purposes by some upon others. Further, certain theories of knowledge that concentrate on 'focus' and rationality may also form atmospheres that *prevent* certain types of atmospheric knowing through the potential of direction of attention (Gatt 2013a, 2013b).

On the one hand, the attunement that Stewart, Throop and Shapiro write about is definitely one way in which the notion of atmospheres is very valuable: it draws our attention to how atmospheres allow us everyday knowledge of the way that we are permeated, constituted by what they are composed of. On the other hand, the notion of atmosphere as a medium, as an existential precondition obfuscates the way in which such atmospheres can be deliberately changed. These changes may be well meaning such as the way performers attempt to generate atmospheres through vibrational song. The work and investigations of experimental theatre draw our attention specifically to the way in which atmospheric knowing depends on changing atmospheres, and requires that we ask further questions about others ways in which atmospheres are shaped and by whom.

Notes

1 Similar concepts are the 'living body' (Farquhar and Lock 2007) or the 'body in life' (Magnat 2014), I discuss these below.

2 I would like to thank Gey Pin Ang for her generous participation in our collaboration, Tim Ingold and other members and associates of the Knowing from the Inside project, University of Aberdeen, the ERC for funding the project, all the participants in Sourcing Worksessions who also participated in my experiments, the Inter Faith Centre at the University of Aberdeen for generously allowing me to use their hall as my studio, even though I couldn't explain very well what I would be doing.

3 www.sourcingwithin.org

4 "One can be a chorus even if one is an individual, each one of us has a thousand voices" (author's translation).

5 As the theatre critic and philosopher Jacek Dobrowolski notes, vibrational singing is "not the most fortuitous phrase, since all singing, as all sound, is vibrational in character, but let's say that what is meant is a specific vibrational frequency of special intimate intensity" (www.taraka.pl/jerzy_grotowskis_dancing_saviour).

6 I participated in six work sessions, in Avellino May 2013, Manchester August 2013, Lisbon January 2014, Barcelona January 2014, Aberdeen March 2014, Aberdeen May 2016. The two sessions held in Aberdeen had a slightly different structure to Ang's work sessions as they were planned with anthropologists in mind as well as performers.

7 Spatz (2015) refers to experimental theatre as embodied research. Precisely because of the aim to transcend questions of the body as separate from other aspects of life, such as either mind or spirit or energy, I find the term 'embodied' problematic and for this reason do not use it.

8 www.marcpetitjean.fr/films/maud-robart-la-source-du-chant/ accessed 15 June 2016 "It moves, it moves things, it moves forms of thought, it moves emotions, it moves something in the body, so that when one is asleep at night the work continues in one's sleep. The body continues to dance in one's bed, because they are songs that wake something up, they call up something that we could call still instinctive. So this continues, the human person in their totality is touched, moved". (Author's own translation, any errors or misinterpretations are my own).

9 I am grateful to the editors for having highlighted that this critique does not fairly represent the approach in the untranslated writings of Böhme. However, Ingold's point, and here this resonates with my argument too, is that much humanistic writing on atmosphere does not attend the particular material qualities of what is referred to as atmosphere. The notion atmosphere tends to become a catchall phrase to speak about these qualities without going into detail about how different material atmospheres affect perception and experience.

10 See Meulemans 2017 for a full-length anthropological treatise on soil, chapter 4 focuses on earthworms.
11 The underlying problem here is possibly that this view makes research methods highly problematic. Where does one start if all things are crucial for understanding the emergence of a situation? See Gatt 2010 and 2013a, 2013b for developments of this point.

References

Ang, G. 2017. *Sourcing Within: A Reflexive Investigation of a Creative Path*. Unpublished PhD thesis, University of Kent.

Artaud, A. [1964] 2010. *Theatre and Its double*. London: One World Classics.

Camilleri, F. 2013a. Between Laboratory and Institution: Practice as Research in No Man's Land. *The Drama Review* 57(1), 156–166.

Csordas, T. 1994. Introduction: The Body as Representation and Being-in-the-World. In *Embodiment and Experience: The Existential Grounds for Culture*, edited by T. Csordas. Cambridge: Cambridge University Press.

Downey, G. 2007. Seeing With a "Sideways Glance": Visuomotor "Knowing" and the Plasticity of Perception. In *Ways of Knowing: New Approaches in the Anthropology of Knowledge and Learning*, edited by Mark Harris. New York and Oxford: Berghahn Books.

Farnell, B. 2000. Getting Out of the Habitus. *The Journal of the Royal Anthropological Institute* 6(3), 397–418.

Farquhar, J. and Lock, M. 2007. *Beyond the Body Proper: Reading the Anthropology of Material Life*. Durham, NC and London: Duke University Press.

Gatt, C. 2009. Emplacement in Multi-Sited Practice/Theory. In *Multi-Sited Ethnography: Theory, Praxis and Locality in Contemporary Research*, edited by M. A. Falzon. Burlington: Ashgate Publishing Ltd.

Gatt, C. 2010. Serial Closure: Generative Reflexivity and Restoring Confidence in/of Anthropologists. In *Unquiet Pasts: Risk Society, Lived Cultural Heritage and Re-Designing Reflexivity*, edited by S. Keorner and I. Russel. Burlington: Ashgate Publishing Ltd.

Gatt, C. 2011. *By Way of Theatre: Design Anthropology and the Exploration of Human Possibilities*. Conference Proceedings 'Participatory Innovation Conference', University of Southern Denmark. Available at http://spirewire.sdu.dk/PINC-proceedings-web.pdf

Gatt, C. 2013a. Vectors, Direction of Attention and Unprotected Backs: Re-Specifying Relations in Anthropology. *Anthropological Theory* 13(4), 347–369.

Gatt, C. 2013b. Enlivening the Supra-Personal Entity: Vectors at Work in a Transnational Environmentalist Federation. In the forthcoming special edition of *Anthropology in Action*, 'Inside Organisations/Outside Structure and Agency', edited by V. Peacock and P. Kao 20(2).

Gatt, C. 2015. The Anthropologist as Member of the Ensemble: Anthropological Experiments With Theatre Makers. In *Theatre as Change: The Transformative Potential of Performance*, edited by A. Flynn and J. Tinius. New York: Palgrave.

Grotowski, J. 1968. *Towards a Poor Theatre*. London: Methuen.

Grotowski, J. 1997. Tue es le fils de quelqu'un. In *The Grotowski Sourcebook*, edited by L. Wolford and R. Schechner. London and New York: Routledge.

Haraway, D. 1991. *Simians, Cyborgs and Women: The Reinvention of Nature*. Free Association Books Ltd.

Haraway, D. 2003. *The Companion Species Manifesto: Dogs, People, and Significant Otherness*. Chicago: Prickly Paradigm Press.
Helmreich, S. 2010. Listening Against Soundscapes. *Anthropology News* 51(10), 10.
Home-Cook, G. 2015. *Theatre and Aural Attention: Stretching Ourselves*. London and New York: Palgrave Macmillan.
Hsu, E. 2007. The Experience of Wind in Early and Medieval Chinese Medicine. *The Journal of the Royal Anthropological Institute* 13, S117–S134.
Hsu, E. 2009. Outward Form (*xing* 形) and Inward qi 氣: The 'Sentimental Body' in Early Chinese Medicine. *Early China* 32, 103–124.
Ingold, T. 1993. The Art of Translation in a Continuous World. In *Beyond Boundaries: Understanding, Translation and Anthropological Discourse*, edited by G. Pálsson. Oxford: Berg.
Ingold, T. 2000a. *The Perception of the Environment: Essays in Livelihood, Dwelling and Skill*. London and New York: Routledge.
Ingold, T. 2000b. Concluding Comment. In *Negotiating Nature: Culture, Power and Environmental Argument*, edited by A. Hornborg and G. Palsson. Lund: Lund University Press.
Ingold, T. 2007. *Lines: A Brief History*. London and New York: Routledge.
Ingold, T. 2008a. Bindings Against Boundaries: Entanglements of Life in an Open World. *Environment and Planning A* 40(8), 1796–1810.
Ingold, T. 2008b. When ANT Meets SPIDER: Social Theory for Arthropods. In *Material Agency: Towards a Non-Anthropocentric Approach*, edited by C. Knappett and L. Malafouris. New York: Springer.
Ingold, T. 2011. *Being Alive: Essays on Movement, Knowledge and Description*. London and New York: Routledge.
Ingold, T. 2015. *The Life of Lines*. London and New York: Routledge.
Irigary, L. 1999. *The Forgetting of Air in Martin Heidegger*. London: The Athlone Press.
Kuriyama, S. 1999. *The Expressiveness of the Body: And the Divergence of Greek and Chinese Medicine*. New York: Zone Books.
Lembo, V. 2016. *La Periferia non esiste: Performance e potere nei progetti teatrali a Scampia, Napoli*. Unpublished Masters thesis, University of Ca' Foscari, Venice.
Maccagno, P. 2016. *Feldenkrais Method: Awareness Through Movement*. Unpublished presentation given at 'Performance Reflexivity, Intentionality and Collaboration: A Sourcing Within and Anthropology KFI Workshop', Aberdeen, May.
Magnat, V. 2015. *Grotowski, Women and Contemporary Performance: Meetings With Remarkable Women*. New York and London: Routledge.
Massumi, B. 1995. The Autonomy of Affect. *Cultural Critique* 31, 83–109.
Merlin, B. 2015. Preface to Bloomsbury Revelations Edition. In *An Actor Prepares*, edited by C. Stanislawski. London; New York: Bloomsbury.
Meulemans, G. 2017. *The Lure of Pedogenesis: An Anthropological Foray Into Making Urban Soils in Contemporary France*. Unpublished PhD thesis, University of Aberdeen.
Nair, S. 2007. *Restoration of Breath: Consciounous and Performance*. Amsterdam and New York: Rodopi.
Rapport, N. 2003. *I am Dynamite: An Alternative Anthropology of Power*. London and New York: Routledge.
Richards, T. 1995. *At Work With Grotowski on Physical Actions*. London and New York: Routledge.
Schutz, A. 1969. *The Phenomenology of the Social World*. Evanston: North-Western University Press.

Shapiro, N. 2015. Attuning to the Chemosphere: Domestic Formaldehyde, Bodily Reasoning, and the Chemical Sublime. *Cultural Anthropology* 30(3), 368–393.

Sloterdijk, P. 2009. *Terror From the Air*. Los Angeles, CA: MIT Press.

Spatz, B. 2015. *What a Body Can Do: Technique as Knowledge, Practice as Research*. London and New York: Routledge.

Stewart, K. 2011. Atmospheric Attunements. *Environment and Planning D: Society and Space* 29, 445–453.

Tatinge Nascimento, C. 2010. *Crossing Cultural Borders Through the Actor's Work: Foreign Bodies of Knowledge*. London and New York: Routledge.

Throop, J. 2014. Moral Moods. *ETHOS* 42(1), 65–83.

Throop, J. 2015. Ambivalent Happiness and Virtuous Suffering. *Hau: Journal of Ethnographic Theory* 5(3), 45–68.

Throop, J. 2016. *Meteorological Moods and Atmospheric Attunements*. Unpublished paper presented at the 'Ethics, Politics, Ontologies Conference', University of Amsterdam, January.

Wolford, L. 1997. 'Introduction to part III Objective Drama', in Wolford, L. and Schechner, R. (eds), The Grotowski Sourcebook. London; New York: Routledge.

Yoeli-Tlalim, R. 2010. Tibetan 'Wind' and 'Wind' Illnesses: Towards a Multicultural Approach to Health and Illness. *Studies in History and Philosophy of Biological and Biomedical Sciences* 41, 318–324.

Zigon, J. 2014. Attunement and Fidelity: Two Ontological Conditions for Morally Being-in-the-World. *Ethos* 42(1), 16–30.

11 The harsh smell of scentless art

On the synaesthetic gesture of hospital atmosphere

Anette Stenslund

> I wonder how nurses and doctors can work in a hospital. You know, the atmosphere is very uncomfortable in the hospital. In there [tilting her head back towards the art installation], you're constantly exposed to flashbacks of such a place. It's very quiet, it smells of hospital, and then there's this public lighting: a killer-light.
>
> (Idun, exhibition visitor)

Introduction

Copenhagen in September 2014. At the National Gallery of Denmark, an extensive exhibition by the Scandinavian artist duo Elmgreen & Dragset opened.[1] *Biography*, as the show was titled, was the second in a series, which had started earlier the same year in Oslo at the Astrup Fearnley Museum of Modern Art. One section of the show consisted of a long labyrinthine corridor (see Figure 11.1). It featured a new constellation of several pieces previously displayed, for instance a piece resembling a waiting room, *It's the Small Things in Life that Really Matter, Blah, Blah, Blah* (2006); two porcelain washbowls with interlacing tubing, *Marriage* (2004); a morgue, *Untitled* (2011); and seven of the numerous doors from *Powerless Structures* (2000–2002). The corridor that brought these pieces together was untitled. From here on, I simply refer to this installation as 'the corridor'.

At the opening, wandering through the corridor, I noted that many other visitors, in subdued conversation with each other, seemed preoccupied with personal experiences of hospitals evoked by the installation. Coincidentally, at that time I had just finished a research project on experiences of hospital atmosphere 'on site' (see Stenslund 2015a), and since this apparent overlap of experienced hospital atmosphere *in situ* and *ex situ* – in the hospital and museum respectively – was too interesting to pass up, I decided to do a small-scale survey. Of interest to me was how the differently situated atmospheres could be related in experience. Eager to learn about museumgoers' ways of addressing the atmosphere of the corridor, I met exhibition visitors just outside the installation in the museum's vestibule and engaged them in conversations about their impressions of its atmosphere. I did

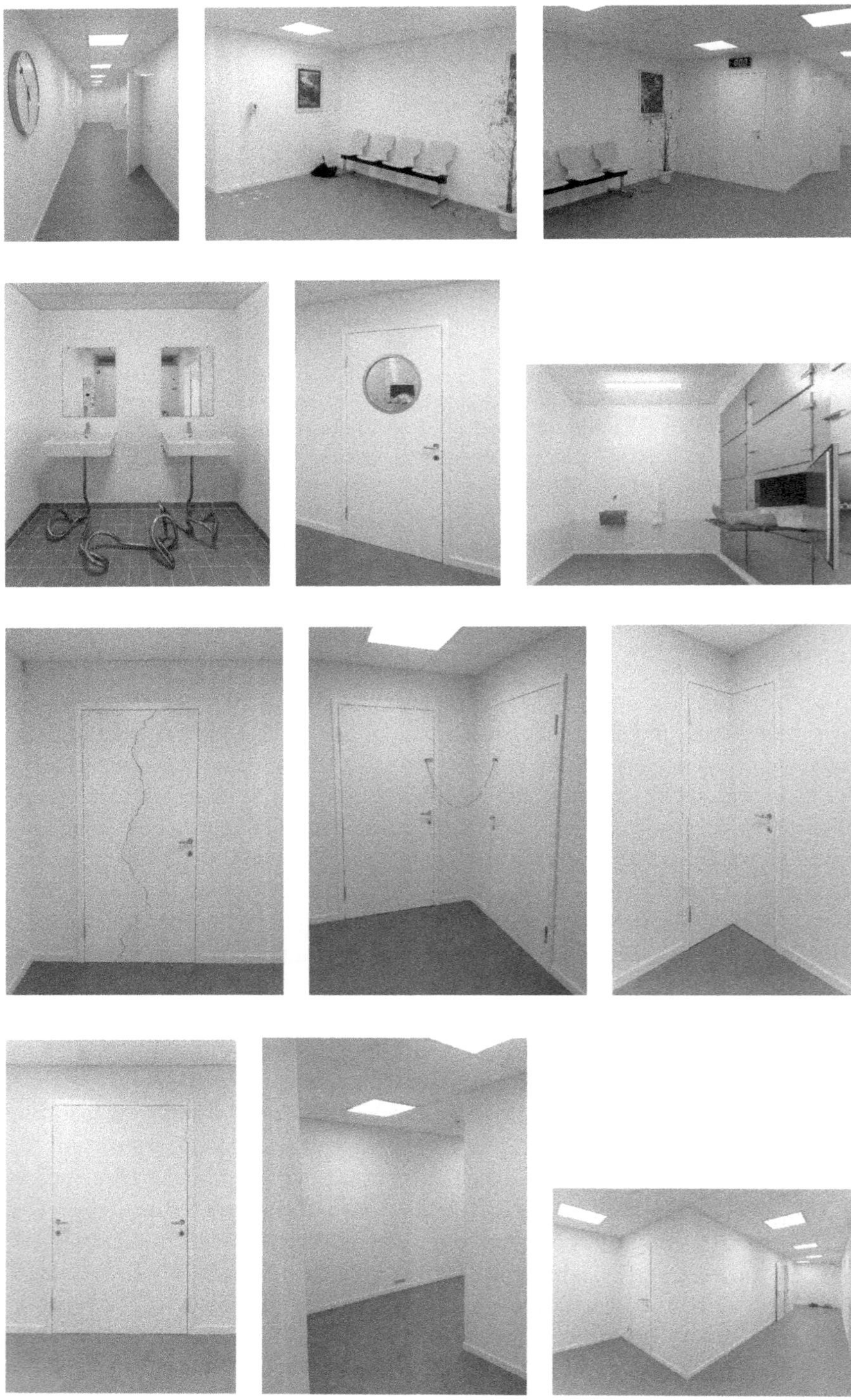

Figure 11.1 *Biography* exhibition (National Gallery of Denmark, 2014)

Images courtesy of National Gallery of Denmark. Photos by Anders Sune Berg and Anette Stenslund roughly a 150 short-length informal interviews in English, German or Danish depending on my interlocutor's mother tongue.

Over the course of the interviews, something strange and unexpected emerged: When I asked visitors about their initial impressions of the corridor's atmosphere – and please note that I did not enquire about visitors' sensory impressions – many began by spontaneously making a direct link between its atmosphere and their smell impressions. In fact, more than 20 of the 150 informants spoke of smell sensations immediately and without prompting. However, no smell was altered curatorially. Intentionally, the artists had not given the corridor as a whole any clear site-specific anchoring to a hospital – the exhibition folder revealed a curatorial motive to evoke a general sense of "a winding, impossibly complex version of a public institution" (SMK 2014: n.pag.). Neither was the corridor supplied with olfactory effects: "Smell is of no crucial importance to us", Elmgreen & Dragset responded to my enquiry.[2] Irrespective of artistic intentions, however, when visitors passed through the corridor many told about a striking resemblance between several of its areas and those found within a hospital. This chapter focuses exclusively on the general experience of a hospital atmosphere, which to many exhibition visitors, as we are to learn, appeared remarkably lifelike – and smelly.

In the following I will explore the question: How is the atmosphere of a *scentless* work of art experienced in terms of *smell*? Atmosphere, here, I define as the affective corporeal involvement in the environment, and I will argue that atmosphere, in this case, is evoked through synaesthetic gestures: through the conjoined interplay of the senses the museum-staged artwork gestured towards hospital atmosphere, hereby connecting institutional provinces that traditionally have been considered far apart due to their dissimilar aesthetic appearances and storied accounts of atmosphere.

For the analytical discussion I draw on mainly two philosophical disciplines: a phenomenological perspective is developed with inspiration from Hubert Tellenbach (1968), Hermann Schmitz (2011), Gernot Böhme (1995, 2001, 2006, 2013) and Martin Seel (2005), and a second perspective builds on inspiration found in the philosophy of language by Ludwig Wittgenstein (1953) and Niels Albertsen (2000, 2012). The conceptual discussion is supported by sensory studies within branches of sociology and anthropology plus ethnographic fieldwork in a hospital and interviews conducted in a museum (Stenslund 2015b). This chapter gives weight to the museum setting.

Beyond my theoretical proposal, I seek to create a space for the empirical findings to come forward, valuable as they are in their own right. So my hope is that the reader will find within the next pages not only a stimulating conceptual discussion, but also invigorating thick descriptions of experienced atmosphere presented in visitor's own wording. I begin by presenting the empirical material followed by a brief introduction of phenomenological work on smell, atmosphere and atmospheric smell. In the proceeding section I take six analytical steps: First, I consider how olfactory sensations of the corridor were tied to biographic life trajectories of visitors evoking a general trace of hospital atmosphere. Second, I debate the felt similarity of dissimilar sensation by pointing to the aesthetic attitude nurtured by the museum context. I thirdly elaborate on the raised aesthetic awareness in the museum that allows for the atmospheric similarity of the two settings – that of the hospital and the museum respectively. This happens in spite of, and perhaps even on account of, the discrepancy between the sensory appearances of the two sites. Fourth, I argue that the possibility of strong smell sensations are generated by nonolfactory sensation. I render possible the nonolfactory experience drawing on selected basic ideas in gestalt theory pinpointing the holistic way of perceiving the environment. Finally, in the last two sections, I open for the discussion on the synaesthetic gesture of atmosphere in the museum holding the potential of moving beyond the separation of senses towards synaesthesia in lived experience.

Interviewing and data organisation

The interviews conducted in *Biography* were vaguely structured allowing the visitors to set the agenda to the widest degree. This was done in order to nurture an experience-near first-person account of the lifeworld against which each informant meets the world (Brykczynski and Benner 2010: 115). Throughout the interviews the only two consistent questions ensuring the themes of discussion were: 1) an opening question querying the visitor's experiences of the corridor's atmosphere; and, if they did not touch on their sensory experiences unprompted, 2) a closing question as to whether they had gained any particular sensory impressions.

I would allow informants to address the question of atmosphere in terms of self-imposed subjects of conversation, and in this way I sought not to provoke a "sensory turn" from the start. Insofar as was possible, I followed up with clarifying questions on subjects broached by the visitors themselves, and I aspired to query about personal past experiences that could possibly colour each particular visitor-experience of the corridor. Since I was not working from a hypothesis about olfactory experiences being of any extraordinary significance to museumgoers, I was not particularly worried about provoking olfactory bias of sorts. The direct way in which some visitors addressed the question of atmosphere with reference to smell came as a genuine surprise to me.

I made brief notes on all of the interviews during the conversations, then took time after each one to write down any quotes that I wished to remember verbatim. I conducted around 30 interviews at a time and transcribed all of them into English directly thereafter. Sorting out the material was both concept-driven and data-driven. First, while going through the material looking for patterns, I excluded accounts not relating to hospital experiences. Also, I separated out more educated interpretations of art and social criticism guided by theoretical reasoning of well-read visitors. This first selection of portions of the material allowed for a phenomenological preceding of mine interested in describing visitors' ways of engaging in their surroundings from a 'lived perspective' rather than investigating their theoretical stances (Giorgi 1997: 236; Bevan 2014: 136). Second, I sought to group conversations according to a differentiated classification of all sensory perceptions mentioned by visitors including temperature, light, sound, smell, touch, kinesthesis (sensation of muscular movement), and sense of time and place. Subsequently, however, it occurred to me that 'carving up' sensory perception as I was in the process of doing seemed too brutal a way of forcing conceptual differentiations onto the material, for often senses were mentioned in a jumble demonstrating more of a sensorial interplay than a neat analytical separateness. Thus, I dissolved the categorisation in order to not split up experientially conjoined sensations. Then I placed a focus on smell impressions for the reason already mentioned – the strange occurrence of smell in a scentless art installation – although, as we are to learn, this focus would not exclude other types of sensation.

In the following I refer to scent, smell and odour – and the lack of the same – interchangeably. Aiming at a 'realist reading' (Brinkmann and Kvale 2014: 270) assuming to observe informants' points of view, another thoroughgoing subject seemed to occupy exhibition visitors: the sense of realness, that is, the realistic yet not 'really real' impression of the atmosphere in the corridor, as one visitor suggested. This data-driven topic I therefore included by allowing it to organise great parts of the discussion. And as I unfold the empirical material, sense perception and atmospheric sensing respectively will be discussed in terms of visitors' sense of realness. I elaborate on this point already in the next section, where I present the relationship between smell and atmosphere and atmospheric smell within phenomenology.

Atmospheric smell

What initiated the upsurge in interest for the concept of atmosphere in phenomenology – highlighting "the emotional tinge of a space" (cf. Böhme 2013: 2) – seems to derive from working on smell. The first to tie in smell with atmosphere was the Heidegger-inspired psychiatrist Hubert Tellenbach who in 1968 would take smell and atmosphere to be one and the same thing:

> The smell *is* at the same time atmospheric. *It concerns simultaneousness.* [. . .] in no other sensuous experience it becomes as clear, that beyond the, in a narrow sense, purely sensuous, something conveys the essence, which stems from the smelling.
>
> (Tellenbach 1968: 46, emphasis in original, my translation)[3]

I suggest considering the 'purely sensuous' smell as a merely perceptual matter. To be precise: the chemical molecular stimuli and the olfactory anatomical and neurological predisposition of processing odour information. Beyond this biochemical constitution of smell, Tellenbach urges to explore smell phenomenologically, that is, not through scientific definition but through a perspective which tunes in on how smell appears in experience. In experience, smell touches our state of being (*Befindlichkeit*): it exceeds 'itself' and becomes part of us.

Gernot Böhme argues that smells are essential components (*ein wesentliches Element*) of the atmosphere; perhaps *the* most essential part, "because smells are like no other sensory phenomenon atmospheric", Böhme amends (2006: 128). Even if I reserve the right to stay reluctant to statements claiming the universal superiority of one sense over other senses – don't forget what we learn from sensory studies, teaching us in detail how different ways of sensing are always situated and culturally dependent (see e.g. Howes and Classen 2014) – there is no doubt about the evocative power of smell in many situations. Smell may often pay a crucial role in how the atmosphere of a given context is experienced; this is so even in a very subtle way without our awareness. Not least this is why smell is widely employed as a marketing strategy and why scent branding has turned into a multibillion-dollar industry (see eg. Lipman 1990; Gulas and Bloch 1995).

Tellenbach collates the human 'state of being' with a 'state of feeling' hereby paving the way to an olfactory phenomenology from which to carry out closer investigations of what I term 'atmospheric smell', defined as smell that marks moodwise or, with Tellenbach's wording: "[d]ie Duftcharaktere, von denen wir uns *Stimmen* lassen" (1968: 35). Albeit Tellenbach holds a specific interest in 'the smell of the nest' – or what he also calls "the aura of family atmosphere" evoking feelings of confidence and homeliness (1981: 229). Aside from such acclimatised atmospheres, he mentions that along with other institutions "hospitals [. . .] have their typical smell and with it their specific atmosphere which reveals their characteristic qualities more comprehensively than does their physical equipment perceived by higher sense-organs" (1981: 228).[4]

Henceforward I am prompted to consider the intertwined relationship between the sensuous environment and the perceiving person, the museumgoer at *Biography*, more closely by way of phenomenology: With Tellenbach I seek to intensify my sensitivity towards the "*kinds of relationships exist*[ing] *between the environment and the self, and how they are constituted by certain situations induced through smelling . . .*". (1981: 222, emphasis in original). Often, I opt for the term 'sensation' to describe the affective, atmospheric surplus of 'pure' smell perception.

Yours, and that of everyone else

"Excuse me, may I ask about your impression of the atmosphere of the installation you've just left?" This was how I would usually address visitors on their way out of the corridor. Deliberately I broached the word 'atmosphere': indeterminate, diffuse and vague as it may appear in aesthetic and political discourse, I sought to embrace the variety of much more exact expressions of atmospheric characteristics that presumably flourish in everyday speech (Böhme 1993: 113). Accordingly, I refrained from specifying what 'atmosphere' would mean to *me*; it was visitors' versions that I was after. Only at rare intervals, if informants were asking for specification, I would refer to the 'mood' roused by the corridor. This way of 'bracketing' my theoretical informed working concept of atmosphere is congruent with phenomenological methods of interviewing found in works by e.g. Giorgi (1997), Bevan (2014), and Brinkman and Kvale (2014). The advantage of this approach, I believe, is to become better at embarking on the myriad of different ways the phenomenon of interest may appear to each person interviewed.

Ruth and Lisa, two exhibition visitors, provide with an example of how personal remembrances of a hospital *in situ* could be evoked by the corridor:

I[5]: *Excuse me, may I ask about your impression of the atmosphere of the installation you've just left?*

RUTH: It smells bad.

I: *How does it smell?*

RUTH: I don't know, it just smells bad. Nor do I understand the purpose of this exhibition. Why do they bring into the museum such kind of stuff? It's a piece of insignificant everyday triviality. We all know it so well. It's not surprising at all.

I: *Are you disappointed?*

RUTH: I am. It might be great art, but it doesn't do anything for me.

I: *It's interesting what you tell me about smell. Could you elaborate a bit?*

RUTH: I don't know, but it smells like in the hospital where my mother's ex just stayed.

LISA [TO RUTH]: Is it perhaps alcohol or some kind of disinfectant?

RUTH: No. Maybe it's actually exhaled from the visitors themselves. Perhaps they emit an odour evoked by the anxiety they experience while passing their way through the corridor. So this is how they react anxiously, you know. It smells like when animals excrete secretions.

I've been in the hospital twice this year as a relative. My mother also died this year. [. . .] It can also be my own associations, of course. Or people in fact really held bad breath [in the installation]. I think they held bad breaths [. . .] it was like in the hospital.

It may be that I begin to associate what I myself experienced lately. In that case, it would be me bringing forth the smell. Did they install smell?

I: [Smiling] *That I don't know.*

Around the time of this interview I was still waiting for Elmgreen & Dragset's confirmation about possible alterations of smell in the exhibition. Their response, when it came, read as follows: "When making immersive installations like these we consider every element, and consciously add or remove sensory expressions as a key part of the overall atmosphere of the work. Smell is not a crucial element for us" (email to this author). Certainly, no place is ever truly scentless, but, again, in the corridor there was definitely an absence of built-in scent effects.

As the exhibition visitors mentioned so many smells – sterility, stuffiness, bad breath, filth, cleanliness, clinic, doctor, rubber, alcohol, cucumber, disinfectants, death, and hospital – obviously not all informants smelled the same thing. Yet the installation seemed to evoke the same site-specific atmosphere:

PETE: I get a weight on the chest. . . . It really makes associations spring to mind. It smells no different than out here [in the museum vestibule], but you feel it smells stuffy and sterile in there [in the corridor]. It's quite crazy and sensational that you are taken directly to a hospital.

Pete's smell experience was different from Ruth's experience. It was stuffy and sterile, and with no bad breath or anxiety-provoked human excretion, yet Pete was also "taken directly to a hospital". Often this transit in time and place marked visitors bodily and mood-wise. Pete felt a weight on his chest and in the below case of Ellen and Selma we see a clear connection of smell, memory, mood and bodily sensations:

ELLEN: It made me feel sick. Honestly. I still feel bad mentally.
I: *How did you get sick?*
ELLEN: Well, it really just smells bad from the moment you enter, so I couldn't help it.
SELMA [TO ELLEN]: Yeah, and because you were taken ill I sought to take it with an effortless ease as a way to compensate, I guess. You didn't say a thing. I did the babble.
I: *How did it smell?*
ELLEN: Like death. It's a clean smell. I know it, because once I worked with dead people.
I: *Where was that?*
ELLEN: I was once doing an internship as a nurse in a hospital.

It came out of these visitor conversations in no uncertain terms that each visitor had his or her own personal version of what characterised the odorous atmosphere. The corridor triggered smell memories of particular hospital experiences – memories that several 18th-century French men of letters acquainted us with before Marcel Proust was even born, although it was his name nevertheless that became associated with them as 'the Proust effect' (van Campen 2014: 20). Thus, owing to such smell-evoked memories, at *Biography* visitors easily got to feel the presence of past *in situ* hospital atmospheres; for Ellen, apparently, the corridor traced back to memories of the clean smell of an internship with dead people.

In the conversation with Ruth and Lisa, one of the first things Ruth conveyed was her disappointment about the familiarity of the scenery on display: "We all know it", she exclaimed. To her the corridor was "a piece of insignificant triviality" and thus "not surprising at all". And then, on my request, she returned to the subject of smell, to elaborate: "it smells like in the hospital". Specifically, it smelled of the two hospitals where her mother and her mother's ex had stayed, respectively. Ruth's experience-based portrayal of the odourous atmosphere suggests that she knew this smell tremendously well. But which kind of smell was it, then? First, Ruth seemed not completely capable of clarifying it to herself: "I don't know, it just smells bad". Later, however, she addressed the smell again, more convinced of its source: 'bad breath' and anxiety-provoked excretions of body odour, just "like in the hospital" she recently experienced. "We all know it so well", Ruth said. A rhetorical question is pressing as to whether it was in fact she, and not we, who knew this smell so well? Seemingly her friend Lisa had got it all wrong by suggesting the smell of alcohol or disinfectants – proposals Ruth brushed off with a clear "No".

Whether smell was materially present or not, visitors would smell the corridor according to personal experiences, allowing them to engage with and become part of the installation. In this sense, one could conceive of *Biography* as having an autobiographical streak brought along not through artist intentions but through smell memories of the visitors. "Whose biography is it?" director of the Astrup

Fearnley Museum of Modern Art Gunner B. Kvaran once asked Elmgreen & Dragset in an interview. "Yours. And that of everyone else", Ingar Dragset unhesitatingly countered (Kvaran 2014: 39–40). Quite so. A wide range of idiosyncratic smell sensations, each characterised in peculiar terms, still managed to bring about a common trace of hospital. There was a corresponding experience of hospital atmosphere irrespective of sensory variations and olfactory divergences.

The atmosphere is right, not sensuously real

How is it that visitor-experiences matched up a smelly hospital atmosphere, when visitors' ways of describing its characteristics did not correspond? In order to look into this question, I now turn to the discrepancy between the factual sensory environment and the actual atmospheric sensation tolerated and even acclaimed in the aesthetic awareness of visitants to *Biography* nurtured by the museum context.

No exhibition visitor really believed that they had wound up in a hospital, yet many were left with a clear sense of a hospital atmosphere that was 'very lifelike', incredibly real, but not actually real, as suggested, for instance, in the following excerpt from a conversation with Anton. He expresses his awareness of the 'otherness' of the museum-staged atmosphere while simultaneously feeling convinced about its 'rightness' in hinting at a hospital:

ANTON: It's very recognisable. A hospital. You are not quite there, but then again you are there. It obviously triggers what you expect to find and recognise. I felt like recognising the smell of the hospital, but I'm not sure if it was real, it's just the whole atmosphere that makes it. You know it all. You know exactly how it feels, which is why you don't feel any need to touch to make sure. This is the case, for example, with the linoleum floor. You don't feel it's necessary to physically test it, for it's just so right. It's another place. One doesn't expect to be in the real place. You know very well that you're not in the real place, but then again it's just so weird. There's a kind of play going on in there, a play with expectations. You know you're not there, but then again you're there.

Anton felt convinced about the rightness of the museum-staged hospital atmosphere. Admittedly it was not a hospital in the sense of 'the real place', but it was 'another place' that felt 'so right' that there was no need for physical testing. This awareness of the *felt* similarity of dissimilar sensations was prevalent:

STORM: The scent is right. It's the institutional smell and also the light fits well. Perhaps, in fact, hospitals are more greenish, but it's the right mood they catch.

NOAH: Yeah, very realistic, but from time to time you also laugh. It's very simplistic and accessible. . . . I think mainly of a hospital when I walk through the corridor. The clock showing that time is standing still. I've just been hospitalised myself. It's very institutionalised. I couldn't smell anything, but the light was spectacular. It's a well-lit display. The white is very white, almost overdone. It's not that white in reality, but who says it should look alike?

Neither Storm nor Noah expected the *ex situ* hospital atmosphere to be mimicking the 'real thing'. Compared with 'reality', Noah found the whiteness of the 'well-lit' corridor overdone, yet it appeared 'very realistic', and to Storm, 'in fact', hospitals would be more 'greenish', but still the artists had succeeded in capturing 'the right mood'.

With only a few exceptions, in general visitors gave the impression of not expecting to find an exact representation of a hospital setting. They seemed to join in the set-up, embracing its artificial yet lifelike appearance. Accordingly most visitors showed themselves to be aesthetically aware that the *ex situ* museum context was 'right', but not 'really real'. The delicate sense of ambiguity of (dis) similarities between the *in situ* and *ex situ* atmospheres expressed by visitors may offer an experiential expansion cherished by the museum setting: "[T]he discovery of what is in truth the case does not, in the aesthetic context, necessarily bring about a devaluation of the preceding perception. It can be registered as an expansion" (Seel 2005: 60). This aesthetic attitude seemed to be prevailing when visitors showed aware of entering the artificial setting of the corridor and met with open minds its sensuous 'betrayal', misfitting on one level yet fitting on another; "very realistic, but also you laugh", Noah told.

The aesthetic attitude attunes to the gestured atmospheric fit by a sensory way of twisting

Diverse forms of sensation can form a gestural kinship in the sense of Ludwig Wittgenstein's (1953) notion of 'family resemblance', which social scientist Niels Albertsen (2000) collectively designates as belonging to a so-called 'semiosphere', a term coined by Yuri Lotman (1990): "In this semiosphere of gestures we can show our understanding of a piece of music, a picture, a poem, or architecture 'in itself' by pointing towards, by comparing it with *something else*. This is exactly the way in which this 'in itself' is maintained" (Albertsen 2000: 102, emphasis in original). Albertsen's proposal helps to consider how even if the corridor by Elmgreen & Dragset did not mimic a hospital sensuously realistically; it still evoked a 'right' hospital atmosphere. The corridor, in fact, seemed to make a virtue out of being something very different, and by so doing it "felt right", as Storm said. It caught the right mood and hence it allowed visitors to engage aesthetically in the hospital atmosphere in ways that the *in situ* setting would not allow for:

RONJA: They catch the hospital atmosphere very well. In a playful way, by twisting.

AXEL: Right, this is like visiting the hospital. . . . But in the exhibition you're aware of the surreal. It definitely brings back memories, but simultaneously they [the artists] subvert it. They flip it. Within the museum one behaves differently than outside. You're curious and you can joke about it, but in real institutions you know it's just shit.

I: *Did you notice any sensuous qualities in there?*
RONJA: I experienced silence, sterility and sadness. It was cold and clinical.
AXEL: There is a flat light. A hospital lighting.
I: *How is it flat?*
AXEL: It's everywhere. Indefinite with no spot and no shadow. Quite fitting. Also the smells were very recognisable, and the lighting was everywhere, yet there was a dark mood.

The artists caught the right atmosphere in "a playful way, by twisting", Ronja suggested, while Axel pointed to the paradoxical way in which the light managed to create a dark mood. It was described as 'quite fitting' hospital lighting, and the smell simply seemed to adapt to the fitting scenery: it was 'very recognisable'. The light was flat, throwing no shadows, infinite and everywhere. Again, visitors were aware of the sensuous dissimilarity of the staged hospital atmosphere compared with a real hospital-based one; nevertheless, it appeared 'just so right' and 'fitting'. Also, Axel noticed the change of attitude from the hospital (where "you know it's just shit") to the museum (where "you can joke about it").

Certainly there are obvious differences between a hospital and a museum and their respective 'inhabitants'. To call on a hospital may be of a vital necessity, one that people would rather do without, whereas a museum visit could be considered a recreational and curiosity-driven leisure activity. Thus a museumgoer's state of mind is liable to be less trouble-minded and burdened by worries, and s/he would probably, unlike a hospital patient, be able to attune to the aesthetic environment more openly and at ease. From Martin Seel we learn that aesthetic attentiveness is a form of awareness where we "take time for the moment" relinquishing a solely functional orientation in life (2005: 20). Admittedly, the aesthetic awareness is not delimited to the domain of art museums, although such places may encourage the sharpened aesthetic awareness particularly well. In light of this assumption not only may museums, but also may art, offer a space relieved from the demanding request for our immediate response, hereby providing us with a momentary pause from the less well-considered and articulated flow of life, from where we can digest and acknowledge the sensory qualities of life in further detail (see Böhme 1995: 17).

The raised sensuous awareness afforded by the museum helps to consider how the felt similarity of a gestured hospital atmosphere was established regardless of dissimilar sensory sensations. That is, even if the relation between the sensory qualities and the overall atmosphere is characterised by incongruity, no one-to-one measure of sensory effect were expected by visitors or even played on artistically.

The nonolfactory gestured gestalt of hospital atmosphere

Even if we buy into the fostered aesthetic attitude of visitors, we may not yet feel compelled to back down on suspicions of visitors' pure illusions: can it be excluded that visitants to *Biography* were not making it all up when bringing to the forefront olfactory experiences of atmosphere? In this section I seek to offer resistance to considerations concerning smell being grasped as pure illusion, and

I seek to render possible that smell in experience may reflect a perceptual way of engaging in the environment assisted by 'nonolfactory' sense qualities illuminated with reference to gestalt theory. For that purpose, I will briefly return to the conversation with Ruth and Lisa presented in the introductory section.

In her attempt to specify her smell experience, after she refused her friend Lisa's proposal about the smell of alcohol and disinfection, Ruth became more self-critical: "It may be that I begin to associate what I myself experienced lately. In that case it would be me calling forth the smell". Ruth seemed to consider the possibility of her having a deceiving sense of smell. Did she realise that something might appear to her that was perhaps not there in reality? Ascertaining that you have perceived something incorrectly can be painful and can motivate you to correct your knowledge (Seel 2005: 60), and just so, in this case, Ruth turned to me for certainty, perhaps as a way to handle her potential delusion: "Did they install smell?"

A prevalent thesis in contemporary brain-imaging studies, first proposed by neuroscientist Jerzy Konorski (1967), suggests that all sense perceptions travel not only from organs to brain, but also from brain to organs, and in terms of olfaction the idea gave rise to theses concerning *phantosmia* being the medical term for olfactory hallucination – a rare olfactory distortion where smell is perceived in a complete physical absence of odorants (Hong et al. 2012; Keller and Malaspina 2013). Elmgreen & Dragset did not install any smell in the corridor; nevertheless, it is patently absurd to have all visitors who sensed a smell of hospital, characterised in various ways by bad breath, death or the like, diagnosed with a rare olfactory distortion or, even worse, to have them excluded from their situated cultural embedding, locked away in figments of a brain. Ruth's sense of smell was not deceiving her and she was not standing aloof from the world in pure imagination either. Contrariwise, I suggest a conception of the scent-sensitive visitors to the corridor as having vast sensations of their surroundings in complete: As culturally informed individuals they likely experienced how other senses – nonolfactory sensations – triggered smell associations and gestured what would liken an olfactory gestalt.

Gestalt, from German meaning form or shape, figures as a model in psychology concerned with the idea that we perceive *whole* constellations of sensory stimuli rather than bits and singled-out parts. Also the conviction consists that we actively co-produce our sensory perception rather than simply react to it (Malnar and Vodvarka 2004: 44). This theory would go far toward explaining the presence of material absent smell by reminding about the complete perception of a partially suggested 'shape' taken off by nonolfactory sensations. Smell needs a 'nonolfactory context' for its completion, Alfred Gell once wisely noted (2006: 402), and at *Biography* it seemed plausible that nonolfactory experiences would make expecting visitors fill in missing 'tunes' on their own, and in so doing strike up the complete overture of a hospital atmosphere. That is, missing parts such as uninstalled smell could become present through their material absences and appear as 'loud smell' of a hospital.

Like many psychologies, gestalt theory also seems to suffer from the unfortunate inclination to reduce almost any kind of perception to the unconscious intrinsic and automatic processing of stimuli, and consequently it fails to recognise the crucial role that also cultural learned associations can play in atmospheric sensation. This is why I bide by pulling out merely an inspiration source from gestalt theory with the ambition to deploy it within a cultural context of skilfully expecting visitors. Yet another weak point of gestalt theory is the tendency to deliver explanations at the expense of factoring in lived experiences. In order to balance this phenomenological neglect below, I seek to pursue the felt sensations of *how* atmospheric smell appeared in experience. This happened exactly, like argued above, along with other senses, and accordingly I seek to vary the discussion of atmosphere through concepts of gesture and syneasthesia that jointly emphasise the complex composite compounds of the atmosphere experienced by visitors to *Biography*.

The atmospheric gesture of the museum exhibition

From Albertsen we learn about the potential for transporting experience of *in situ* atmospheres through the use of gesture (2000, 2012). Gesture, here, is a concept deriving from Wittgenstein that embarks on a semiotic analysis of meaning experience of the use of verbal and non-verbal language, as when, for instance, the tone or the rhythm of words adds something to the felt impression that words may also help to pass along. In Albertsen's wording, gesture is not about "*what* we hear, but *how* we hear" (2012: 73, emphasis in original), and how we hear, see or otherwise perceive words adds a felt experiential dimension to the expression; it becomes atmospheric (Albertsen 2000: 78). It is the merger of sensuousness and meaning – a sense-making happening in the twofold meaning of the term; to sense and to make sense – that can turn language into an atmospheric source owing to its gesture (Albertsen 2012: 71), and by striking up a gestural emotional similarity to "the original" *in situ* atmosphere one can present atmospheres anew without representing an identical clone (Albertsen 2012: 72). Take a good poem; obviously it would not need to drip with water in order to carry its reader away in a felt sensation of water trickling down a rivulet. If it did, it would likely fall flat and turn into a silly gimmick. A good poet in turn would make you *feel* the water through the *use* of language.

Analogous to the atmospheric gesture of language, artwork and exhibitions in general can gesture atmosphere in the sense of making visitors feel it. Exhibitions are certainly not books, and artworks, exhibition designs and collection pieces are not words, yet they communicate with visitors through gesture. From the following quote we get the impression that also smell may be gesturing and that the olfactory gesture is less about *what* we smell but *how* we smell:

I: *How would you describe the atmosphere?*
ELIAS: It's unpleasant due to the smell.
I: *How is it unpleasant?*
ELIAS: It's kind of locked up and claustrophobic like: "Now that you've entered, you'll never get out again". You feel forsaken and powerless. It's very labyrinthine. You're constantly thinking, when does it end? You don't know

what's waiting for you round the next corner. It smells clinical and there's this fluorescent lighting, which also fits into the clinical. Simultaneously it was a used atmosphere.

I: *How was it used?*

ELIAS: It was like, in the waiting room, for example, with all the tickets lying around and the magazines that had been browsed through innumerable times. Somebody was there before you, but also it's very institutional and completely anonymous and with a dead weeping fig. You've been there.

Elias describes the smell of distress about being "locked up". The claustrophobia seemed to conjoin with the labyrinthine architecture of the corridor that evoked feelings of being powerless and forsaken. It was all tied to the smell. Together with fluorescent lighting, a well-worn magazine, tickets lying about, and a dead weeping fig, it all "fit well with the clinical", Elias said. The smell was clinical, too. What made it fit could be a character sketch of a hospital, I suggest, and here, no longer in its particular form tied to autobiographic past experiences only, but also in its indefinite form of what we could term 'hospitalness' being a hospital-characteristic that most late-modern Western museumgoers would be able to agree upon, owing to their cultural embedded common knowledge and bodily ways of knowing.

Presumably it is not hard to imagine that the sensations described by Elias resemble those of patients in a hospital. The disorientation may be both existential and literal. Patients are compelled to place their fate in practitioners' skilled hands, often unable to fully understand the treatment procedure: like Elias, they "don't know what's round the next corner waiting for them" (see, by way of comparison, the top left corner of Figure 11.1 illustrating the physiognomy of the corridor in this respect). Moreover, and slightly more down-to-earth, from my own experience during the ethnographic fieldwork in a hospital, a characteristic feature was long corridors all looking the same, at least to the unaccustomed eye of a patient. Accordingly, on a daily basis patients would turn to me asking for directions. In the same way, the maze-like architecture of the corridor provoked a feeling of losing sense of direction existentially as well as in terms of navigation. Elias also pointed to the tickets and the worn pages of the magazines giving the impression of having to wait – perhaps indefinitely? He felt forsaken and powerless; nobody was there and did anybody care? Not even the weeping fig had been taken care of. It had died, and its dry leaves lay around, just like the used tickets, on the linoleum floor. "You've been there", Elias concluded, suggesting not only that he knew, but also that we would know this Kafkaesque feeling of hospitalness.

The synaesthetic gesture of hospital atmosphere in the museum

At one level we know very well that smell does not exactly lock somebody up, and also it may not be smell or light as such that is clinical or sterile – or that may even kill, as insinuated by Idun in the epigraph of this chapter: "It's very quiet, it smells of hospital, and then there is this public lighting: a killer-light". Clinical and sterility are conditions, and the act of killing or locking somebody up would be considered a crime. In spite of this, it still makes sense. With reference to various forms of sensations in what may at first seem to be a multisensory indexical hotchpotch, Elias provides us with a synaesthetic characterisation of the atmosphere. Synaesthesia literally means "joining of the senses", and the concept plays a crucial role in works on atmosphere suggested by Gernot Böhme, who makes reference to both Aristotle's *sensus communis* and Goethe's sensory-moral effect (*sinnlich-sittliche Wirkung*) of colours, in order to stress, not the moral effect of sensation (which Böhme takes to be a misinterpretation) but the emotional influence colours, among other sense qualities, can have on our "life-feelings" (*Lebensgefühl*) (2001: 97).

Synaesthesia, here, is defined as the prevalent kind of synaesthesia where different sensory qualities exert a mutual influence on each other resembling a cultural associative sensation. The experienced relatedness of different sense qualities appear when people in general tend to speak of light, warm or cold colours; heavy, screaming or calm lighting; round, soft or loud smell; cozy, piercing or crunchy sounds.[6] Here senses conjoin in a value-laden affective unity of impressions that connect with our *Gestimmtheit*, that is the way we are tuned mood-wise. With a view to Böhme's effort in dividing the atmosphere into atmospheric parts – with reference to the possibility of deciphering the situated aesthetics that evokes the atmosphere "[i]t is not sufficient to point out that the whole is more than the parts", Böhme argues (1993: 124) – astounding enough, it is with an expressed wish not to feed an analytical separation of the senses that Böhme states his motive not to apply any other conceptualisation than synaesthesia to describe the conjoined sensation of atmosphere (see Böhme 2001: 90).[7] This might be a wise choice in the sense that what Böhme is after is the *felt* sensation of similarities of senses contained in the concept of atmosphere.

According to conventional reasoning, such comparisons of different sensory matters would easily be brushed aside, yet, as Charles Sanders Peirce remarks with yet another illustrative example, "everyone who has acquired the degree of susceptibility which is requisite in more delicate branches of reasoning . . . will recognise at once so decided a likeness between [say] a luminous and extremely chromatic scarlet . . . and the blare of a trumpet" (Peirce 1932: 156). Different sensory stimuli may strike up the same atmosphere. This notion, enables to fancy the comprehension of the synaesthetic gesture of the art work not being bent on a sensuous *re*presentation that objectively can be assessed as 'good' or 'less good' according to how close it gets to an imitative reproduction of a given original, like, in this case, a 'real' hospital setting. The conceptualisation of synaesthetic gesture emphasises how also visitors enact the atmosphere on show. In brief: the sensuous perception is no neutral act of taking in the surrounding environment; rather it produces and adds an atmospheric layer to the experience. Ontologically, atmospheres can never be separated from the human beings that experience. Atmospheres are defined as phenomena "in-between", and always the perceiving human beings are implicated – we co-produce atmosphere (Böhme 2006: 33), and the current case helps to remind about how this entanglement proceeds in experience.

According to Hermann Schmitz it is the corporeal sensations of synaesthesia that hold together sense-qualities and feelings (2011: 37). And exactly so most of the above-presented depictions of the atmosphere of the corridor rely on visitors' synaesthetic bodily engagements with the display where it makes little sense to single out the senses. In Werner's case, the corporeal sensation of the corridor is particularly evident:

WERNER: You see and smell things that you've experienced ad nauseam. In my 35 years I've been in and out of hospitals, and I get vomiting sensations of walking through the hallway. You're fed up to the back teeth [gesturing with flat hand to neck stop]. It's a mould-breaking exhibition, because it's usually very pleasant to visit this museum . . . but this exhibition is like a thorn in the side of something that is usually very beautiful. It's in your face, with a nasty smell [. . .] like being punched in the stomach.

Werner described the atmosphere as nauseating and nasty smelling, like being punched in the stomach; Elias experienced the unpleasant smell of claustrophobia and powerlessness; Ellen turned decidedly sick from the clean smell of death; Pete felt a weight on his chest and Axel, who also recognised the smell, was marked by the bright light causing a dark mood; Ronja experienced silence, sterility, sadness and coldness; and Idun depicted a clinical smell along with the public killer-lighting. This surely was a direct hit of visitors' resonating bodies that through a criss-cross of various senses felt what was 'in the air'.

Conclusion

The ethnographic working on the synaesthetic gesture of atmosphere, rich in cultural associations and multisensory entanglements and charged with feelings, helps to consider why phenomenologically no sensory quality can successfully be singled out from the atmospheric whole.

According to self-exclaimed visitor experiences of Elmgreen & Dragset's art installation on show at *Biography*, the chapter portrayed the incongruity of the relation between its sensory qualities and the overall atmosphere that it evoked. I have suggested how olfactory sensations of an otherwise scentless corridor were tied to personal remembrances of a hospital atmosphere *in situ*. Sensory effects, I argue, however are not in any clear accordance with the atmosphere, and in this case, the felt similarity of a hospital atmosphere situated in the hospital and the museum respectively, was established regardless of dissimilar and idiosyncratic sensations. Experiences of the corridor were characterised by having a subjective element spurred by autobiographic smell memories, yet the associations were culturally informed and marked by a common denominator tied in with the character sketch of hospitalness and hospital atmosphere. I suggested how the aesthetic attitude nurtured by the museum context could make visitants afford and embrace the ambiguous relation between senses and atmosphere. Further, the chapter elaborated on the possibility of strong smell sensations brought about by the nonolfactory gestured gestalt explaining the experienced presence of absent smell as material matter arguing how partial stimulation can be perceptually completed. Accordingly, the olfactory gestalt opened for a discussion on the synaesthetic gesture of atmosphere emphasising how visitants enacted the hospital atmosphere on show tying together emotional, charged sensations that are often thought of as independent.

It has not been my intention here to claim the superiority of smell over other senses. Contrariwise, as mirrored in many visitor conversations, other senses too – nonolfactory sensation – managed to invade exhibition visitors affectively. Diverse sensory qualities are only dissimilar and disparate in a highly abstract and ordered theorisation, and due to its phenomenological approach the chapter proposes how commonly senses can appear in conjoined constellations and atmospheric wholes; not separately but outright conjoined, fused together in synaesthetic gestures of atmosphere. Hence, the ethnography on atmosphere gets its strength from varying the mainly philosophical accounts on atmosphere so dominant within this field. It has been argued that atmosphere cannot be reduced to sensory parts (see. e.g. Schmitz 2002), but how the relation between senses and atmosphere is in fact reflected literally in involved and engaged bodies and minds that we still know little about. This chapter sought to take a modest step towards greater knowledge concerning the synaesthetic gesture of atmosphere conjoining the museum venue with institutional spheres, *in casu* that of a hospital, outside the museum.[8]

Notes

1 Thanks to the National Gallery of Denmark for giving me access to their venue.
2 My sincere thanks to Elmgreen & Dragset for the correspondence.
3 German original: "Der Duft *ist* zugleich auch dieses Atmosphärisches. *Es handelt sich um Simultanes*. . . . in keiner anderen Erfahrung unserer Sinne wird so deutlich, daß über das im engeren Sinne Vernommene hinaus sich etwas vom Wesen dessen mitteilt, dem das Duftende entstammt".
For clarification I may add that Tellenbach referred to 'atmosphere' and 'atmospheric' interchangeably.
4 The higher and lower ranking of senses seems to be a cultural legacy from Aristotle's hierarchical depiction of the senses with sight and hearing portrayed as the most accurate and delightful senses serving more than nurturing needs (Aristotle 1986 II.8:421a7, see also Synnott 1991: 65).
5 Throughout the chapter all museumgoers are anonymised. 'I' is the interviewer (this author).
6 The 'weak' kind of synaesthesia subscribed to here is not to be confused with the medical term that implies a measurable cross-activation of brain-maps, which is a rather peculiar phenomenon (see e.g. Howes and Classen 2014: 155).
7 The notion of 'synaesthesia' used by Böhme may at first seem to have much in common with to what within *sensory studies* has been termed 'intersensoriality' advanced by anthropologist David Howes (2006). In favour of "the model of synaesthesia", however, Howes suggests to apply the term "intersensoriality", hereby seeking to avoid to bolt together the senses, and thus also to enable investigations to zoom in on sensory *differences* and to focus on the cultural situated values and meanings attached to the way senses may interplay (Howes 2006: 164).

8 Part of the preparation for this chapter was done while working on my doctoral thesis on *Atmospheric Smell* at Medical Museion and the Novo Nordisk Foundation Center for Basic Metabolic Research (CBMR), University of Copenhagen.

References

Albertsen, N. 2000. *The Artwork in the Semiosphere of Gestures: Architecture, Language, Critique: Around Paul Engelmann*. Amsterdam: Rodopi, 67–104.

Albertsen, N. 2012. Gesturing Atmospheres. In *Ambiances in Action: Proceedings of the 2nd International Congress on Ambiances. Canadian Centre for Architecture, Montreal, September 2012*, edited by J.-P. Thibaud and D. Siret. Mayenne: International Ambiances Network, 69–74.

Aristotle. 1986. *De Anima: Books II and II (with passages from book I)*. Translated by D. W. Hamlyn. Oxford: Clarendon Press.

Bevan, M. T. 2014. A Method of Phenomenological Interviewing. *Qualitative Health Research* 24(1), 136–144.

Böhme, G. 1993. Atmosphere as the Fundamental Concept of a New Aesthetics. *Thesis Eleven 36* (1), 113–126.

Böhme, G. 1995. *Atmosphäre: Essays zur neuen Ästhetik*. Frankfurt am Main: Suhrkamp.

Böhme, G. 2001. *Aisthetik. Vorlesungen über Ästhetik als allgemeine Wahrnehmungslehre*. München: Wilhelm Fink Verlag.

Böhme, G. 2006. *Architektur und Atmosphäre*. München: Wilhelm Fink Verlag.

Böhme, G. 2013. The Art of the Stage Set as a Paradigm for an Aesthetics of Atmospheres. *Ambiances. Environnement sensible, architecture et espace urbain*. Online 10 February [Accessed 11 November 2015].

Brinkmann, S. and Kvale, S. 2014. *Interviews: Learning the Craft of Qualitative Research Interviewing*. London: Sage Publication.

Brykczynki, K. A. and Brenner, P. 2010. The Living Tradition of Interpretive Phenomenology. In *Interpretive Phenomenology: Embodiment, Caring, and Ethics in Health and Illness*, edited by K. C. Garrett, K. A. Brykczynski, R. E. Malone and P. Benner. Indianapolis, IN: Sigma Theta Tau International, 113–141.

Gell, A. [1977] 2006. Magic, Perfume, Dream . . . In *The Smell Culture Reader*, edited by J. Drobnik. Oxford: Berg, 400–410.

Giorgi, A. 1997. The Theory, Practice, and Evaluation of the Phenomenological Method as a Qualitative Research Procedure. *Journal of Phenomenological Psychology* 28(2), 235–260.

Gulas, C. S. and Bloch, P. H. 1995. Right Under Our Noses: Ambient Scent and Consumer Responses. *Journal of Business and Psychology* 10(1), 87–98.

Hong, S. C., Holbrook, E. H., Leopold, D. A. and Hummel, T. 2012. Distorted Olfactory Perception: A Systematic Review. *Acta oto-laryngologica* 132(1), 27–31.

Howes, D. 2006. Scent, Sound and Synaesthesia. In *Handbook of material Culture*, edited by C. Tilley, W. Keane, S. Küchler, M. Rowlands and P. Spyer. London: Sage, 161–172.

Howes, D. and Classen, C. 2014. *Ways of Sensing*. New York: Routledge.

Keller, A. and Malaspina, D. 2013. Hidden Consequences of Olfactory Dysfunction: A Patient Report Series. *BMC Ear, Nose and Throat Disorders* 13(8). Available at www.biomedcentral.com/1472-6815/13/8.

Konorski, J. 1967. *Integrative Activity of the Brain: An Interdisciplinary Approach*. Chicago: University of Chicago.

Kvaran, G. B. 2014. Whose Biography? Interview by Gunnar B. Kvaran. In *Biography*, edited by Elmgreen & Dragset and the authors. Berlin: Archive Books, 35–51.

Lipman, J. 1990. Scents That Encourage Buying Couldn't Smell Sweeter to Stores. *Wall Street Journal* 5.

Lotman, Y. M. 1990. *Universe of the Mind: A Semiotic Theory of Culture*. London: I.B. Tauris.

Malnar, J. M. and Vodvarka, F. 2004. *Sensory Design*. Minneapolis, MN: University of Minnesota Press.

Peirce, C. S. [1894] 1932. *Collected Papers, Vol. 1*. Cambridge, MA: Belknap Press of Harvard University Press.

Schmitz, H. 2011. *Der Leib*. Berlin: Walter de Gruyter.

Schmitz, H. [together with G. Marx and A. Moldzio]. 2002. *Begriffene Erfahrung: Beiträge zur antireduktionistischen Phänomenologie*. Rostock: Koch, 13–211.

Seel, M. 2005. *Aesthetics of Appearing*. Stanford, CA: Stanford University Press.

SMK 2014. National Gallery of Denmark: *Elmgreen & Dragset. Biography. 19 September 2014-04 January 2015*. Exhibition folder, Copenhagen: National Gallery of Denmark, n.p.

Stenslund, A. 2015a. A Whiff of Nothing: The Atmospheric Absence of Smell. *The Senses and Society* 10(3), 341–360.

Stenslund, A. 2015b. *Atmospheric Smell: Hospital-Based and Museum-Staged*. Copenhagen: SL Grafik.

Synnott, A. 1991. Puzzling Over the Senses: From Plato to Marx. In *The Varieties of Sensory* Experience, edited by D. Howes. Toronto: University of Toronto Press, 61–76.

Tellenbach, H. 1968. *Geschmack und Atmosphäre*. Salzburg: Otto Müller Verlag.

Tellenbach, H. 1981. Tasting and Smelling-Taste and Atmosphere-Atmosphere and Trust. *Journal of Phenomenological Psychology* 12(2), 221–230.

Van Campen, C. 2014. *The Proust Effect: The Senses as Doorways to Lost Memories*. Oxford: Oxford University Press.

Wittgenstein, L. 1953. *Philosophical Investigations*. London: Blackwell.

12 On the dynamic and duration of atmosphere

Sounding out New Phenomenology through music at China's margins

Friedlind Riedel

New Phenomenology, a school of thought initiated by German philosopher Hermann Schmitz, centres on the critique of the humanist subject and the paradigm of introjection in Western intellectual tradition, after which feelings are identified with a person's inner sphere such as soul or mind.[1] Affected by the dramatic events of World War Two, between the lines of his 10-volume strong *System of Philosophy* published between 1964 and 1980 surfaces the attempt to comprehend the devastating dynamics of the mass support for Nazi Germany that had raised radical questions concerning the seclusion and autonomy of the Enlightenment subject. For Schmitz, such mega events of Nazi speeches or the subversive authority of the *Führer* are evidence to the claim that feelings are not private but collective, feelings that do not belong to singular citizens but that had coalesced into one single "Bloc-*Leib*", a mass (ibid.: Schmitz [1978] 2005: 79, 83). Schmitz draws on spatial metaphors to convey that feelings are not internal states of an individualised subject but, turning the subject inside out, are more appropriately conceived of as atmospheres.[2] Such atmospheres are spatially disseminated but not localisable in spaces (*Räumen*), extended but not measurable in temporal or spatial distances and positions, extensive, holistic, memberless: they cannot be segmented. They are meaningful yet vague, indistinct yet intrusive; they are dynamic stirrings in the region of what is felt as present (*anwesend*). Moreover, rather than surrounding a given subject or mediating between subjects and objects, they elicit and alter modes of existence and mould collectives (Schmitz 2009: 78, 2014a: 85). Music, for Schmitz, exemplifies such claimed "non-subjectivity of feelings" and their impalpability, since music and atmosphere share the same ontological vagueness, are both manifest in suggestions of movement and share the dynamism of contraction and expansion as their very essence (Schmitz [1978] 2005: 260).

The many theories of atmosphere that draw on music and sound phenomena (Connor 2006; Morton 2007; Gumbrecht 2011; Böhme 2013) speak of the kinship of music and atmosphere. Atmosphere in turn may enable us to advance an intellectual framework for researching the multifarious vagueness, the non-fissile relations, the chaotic and irreducible intensity and the unnameable forces that are manifest when music sounds.

In this chapter, I chart some of the incommensurabilities and resonances between Schmitz's notion of atmosphere and the wider field of atmospheric research, by entering into one of the principal disputes regarding the concept of atmosphere: namely the dispute between epistemic holism (*Ganzheit*) in which singular bodies are completely consumed within a diffuse whole; and analytical constellationalism, in which a whole is constituted through the spatial constellation of singular bodies. Put another way, in the holistic approach the difference between relations and relata is obliterated whereas the constellationalist perspective clings to the paradigm of individualised entities and their secondary relations. While Schmitz's notion of atmosphere espouses the first position, the writings of Gernot Böhme (2013) can be identified with the second. In this chapter, I suggest that atmosphere is operative precisely at the threshold, the point of tension, between an indivisible, irreducible whole and the fragmentation, the heterogeneous, the individual person. Ultimately, my concern is not with perceptions of atmosphere and the inevitable ontological questions regarding its substantiality. Rather I aim to explore what atmosphere does, how, and in what way it is a productive force in situations of music-making and listening, and how atmosphere impacts on processes of selves and collectives. I concur with Schmitz in his critic of constellationalism and invoke his notion of situation to point at the limitations of topological thinking preeminent not only in atmospheric research but in sound studies and orthodox musicology (Cook 2014). After all, a rigidly spatial notion of atmosphere paralyses music. Drawing on Schmitz's terminology of *situation*, *dynamism* and *duration*, I aim at advancing a process-oriented concept of atmosphere that is less interested in spatial constellations but rather in performance. This is not to disregard space and the advantage of spatial and spatiotemporal metaphors but to further a musical notion of atmosphere, as has been propounded by Morton (2007) and Ford (2013).

My points of departure are situations of music-making in the village of Manna in the mountainous borderland between Myanmar and China, a sub-tropical region called Bulangshan. Here, the musical everyday consists mostly of the circulation of MP3s played on mobile phones and motorbikes, of local music on VCD or diverse music on TV, and by songs sung solo or as duets, accompanied by a three-stringed lute or a reed aerophone. On top of this, ceremonial music on holidays saturates the entire village with the pervasive sound of drums, gongs, and cymbals. Beginning from mundane music-making and the abundance of feelings of shame and timidity as voiced by my interlocutors when describing their experience in music-making, I ask how the concept of atmosphere may help disclose the dynamics and temporalities of collectives in situations of music-making.[3]

Atmosphere as situation

When I first came to the village Manna I was received at the village square; a plain, concrete place, accessible to everyone, publicly visible. The long flight of stairs that ran from the public square to the school on the hillside above the village was crowded with inquisitive pupils. Teenage mothers, grandfathers as well as those who usually work in the fields at that time of the day had gathered in the shadow of stilt houses. My visit had already been announced beforehand through a middleman from another village, who had been assigned the role of introducing me to Manna, a village, as he said, "full of music". Conscious that the alien person was interested in Manna's music, two men had started making music on the village square upon my arrival. It was early afternoon, and the concreted square was simmering under stinging sunlight, whelmed with swirling heat. While one man beat a drum, another played cymbals, both moving their bodies with restraint, turning their backs to me, while exchanging glances with the onlookers. Another man, whom I would later come to know as AiGianNan, anxiously struggled to stir up the reticent crowd, who gazed at the musicians and myself in motionless silence. As his appeal dissolved in the seeming indifference of the standers-by, he came up with many apologies and explained to me that the unmarried girls who would usually dance were too intimidated to come forward. I felt awkward myself given the apparent diffidence towards me. In a bold attempt to alter the oddity of the situation, four to six men started dancing in a circle, slowly moving arms and legs in sweeping gestures, yet only elicited surreptitious sniggering in the crowd. The drum continued to abortively charge the village with a festive beat, but people did not answer its sonorous call. Even though a few courageous men played instruments and danced on the open square that day, it appeared as if the music was not heard by the villagers. The inviting movements were answered with retreat and hesitation further forging tension. As music grasped at nothing, it finally fell silent. Villagers returned one by one to their routines and I went with a small group of musicians towards the house of AiGianNan. It was only then, inside the house, that the music got going.

This encounter was permeated by a particular atmosphere that structured and governed the unfolding of events. The same sounds and rhythms, the same place and people would, no doubt, operate together in a very different way in another situation, that is, against the backdrop of a different atmosphere. Schmitz emphasises the *situation* as a primary category of his New Phenomenology. In situations the material and the semantic, nature and culture are not separated. Situations instead are characterised by "states of affairs, programmes and problems" (Schmitz 2009: 47) and hence are always particular not to a material setting but are equally shaped by culture and history. Schmitz develops the concepts of "feelings as atmosphere" and situation in such a way that they become almost interchangeable, arguing that most situations are "charged with atmosphere" (1998: 178) and that "feelings are mostly embedded in situations" (2011: 118) Schmitz dismisses the by now prevailing notion of atmosphere as a constellation of things, as proposed by Böhme who insists that atmospheres "proceed from and [are] created by things and their constellations" (Böhme 1993: 122). In contrast, for Schmitz (1998: 185), atmospheres pertain to situations, which he defines in reference to Erich Rothacker as "wholes hanging together by virtue of an internally-diffuse meaningfulness that does not consist of singular numerable elements" (Schmitz 2009: 47). Moreover, this very wholeness and indivisibility of a situation is precisely established by its "atmospheric cohesion" (2005: 46). The originality of Schmitz's phenomenology lies in its radical critique of singularity. According to the axiom of singularity, the world is construed as an aggregate of discrete elements, a network of separate bodies and forces, which by way of their constellation and intermeshing generate an emergent third dimension, an atmosphere. Yet, for Schmitz,

> The world shows up not as a neutral realm of already separate entities but as the atmospheric fields of significant situations, opportunities or quasi-corporeal forces or 'opponents' that in the first instance become manifest to the conscious person in form of the 'internally diffuse meaningfulness' of holistic corporeal impressions. Articulation of significant situations into constellations of separate objects and structures is a later-coming achievement (although it is usually taken as primary by theoretical thinking).
>
> (Schmitz et al. 2011: 244)

The axiom of singularity

With music it is rather evident that singular elements such as pitches, chords, beats, timbre, or background-noise do not in themselves add up to music's immediate intensity, suggestive force, atmospheric character or mood. In anthropology, object-oriented ontology as proposed by Graham Harman (2011) has utilised this approach, in which the very singularity of objects and their inherently non-relational reality operates, to an extreme degree, as a foundational premise.[4] The operative logic of isolating objects or actors from the internally-diffuse meaningfulness of situations hinges on and reinforces singularity and, thus, structurally obscures the atmosphere. As a consequence, the atmospheric dimension of a situation remains effectively unsayable in an object-oriented account.[5] Schmitz's contemporary, the Psychiatrist Hubertus Tellenbach argues in a similar vein, when he writes that the divisibility of a whole as a premise for the production of serious knowledge renders the "scientific study" of the atmosphere impossible (Tellenbach 1968: 60).

The crucial point of Schmitz's phenomenology, however, is that the very possibility to divide a situation or atmosphere into singular objects and networks of elements requires an antecedent involvement in situations (Schmitz 2005: 27). Drawing on gestalt theory of Wolfgang Metzger or Max Wertheimer, atmosphere as situation, for Schmitz, does not "exceed the assembling of bodies" as Ben Anderson (2009: 77) puts it, but precedes the very human ability to identify and differentiate singular bodies in the first place.[6] Here of course Schmitz's atmosphere bears clear resemblance with Martin Heidegger's *Stimmung* which "has always already disclosed being-in-the-world as a whole and first makes possible directing oneself toward something" (Heidegger 2010: 133).[7]

Love

More than merely locating atmosphere in an unsegmented world and asking for a methodology takes into account their indivisibility, I want to approach this aspect from a different angle by saying that atmosphere comes into appearance as an affective force when the singular elements are taken up in an all-encompassing gestalt, a feeling that has lost the differentiability and singularisation associated with language. Yet, music in Manna comes with words. Songs, that narrate the sustained longing for lovers and pasts, the yearning for worlds out of reach. Songs that extol beauty, and speak of jealousy, of marital relations, utopian eternity, fidelity and adultery, but also that allude to death and even bewail the horror of sexual abuse. These various relationship situations are enacted in song by means of a language that draws on the multiverse of the local flora and fauna. AiGianNan sings that he wants to be a tree when he dies. His wife would then be a hornet; she would come and build a nest in the hollow trunk of the tree; she would fly out yet always come back. A mutual encorporation, a collective situation, love. Figurative language, so Morton suggests, is able to "heighten a sense of the radical non-identity of things" (2002: 55). As he performs a utopian rhapsody in song, he sways back and forth, glancing into the far distance of a future further down the cycle of birth and rebirth informed by the memory of a burgeoning love 30 years ago, collapsing time into one meaningful situation to come. Music, Schmitz writes, not only eludes the identification of particular feelings – or better here, emotions – which always seem as a metaphoric approximation to what is really going on in music (Schmitz [1978] 2005: 258). It is precisely this diffuse meaningfulness of the "musical sound" (ibid.: 250) that makes music particularly suitable for elaborating the diffuse yet meaningful feelings that pervade love relationships, into atmospheres. In AiGianNan's song, music and figurative language conflate into one stirring atmosphere of longing.

Situation-listening

In an atmosphere, defined as *situation*, the musical and the extra-musical, the audible and the inaudible, the real and the imagined are indistinguishable. Within atmosphere there exists no music in itself that could be separated from the situation without corrupting the atmosphere to which that music pertains. Hence I have previously proposed that music be conceptualised *as* atmosphere[8] (Riedel 2015). In this perspective, what resounded at the village square that noon in Manna was not festive music in an inappropriate context; it was a meaningful atmosphere that had saturated the situation and that was entangled with, hinged on and that culminated in the dynamics embodied by music. The festive sound of drums and cymbals and the energetic drive in the rhythmic pattern seemed somehow preposterous as the suggestions of movement within the rhythmic refrains were unanswered by the crowd, and as the atmosphere remained tense. Or was it the tense atmosphere that inhibited the rhythmic movement of the felt-body as urged by the drums? The presence of the festive rhythms sounded the absence of a festive atmosphere. Awkwardness was *audible*. The mode of listening that corresponds to the atmosphere is, Schmitz suggests, a situation-listening (*Situationshören*), the hearing of sounds buried in a situation, more precisely: the hearing of a situation condensed in sound or music (Schmitz [1978] 2005: 252).

As the atmosphere is manifest as *situation* it is indivisible into sensorial regimes, so Schmitz tells us, and its character is rather "synaesthetic" (Schmitz 2009: 40). The respective sensations that are eliminated as auditive sensations are, as Tellenbach argues "always already fundamentally moulded by means of the atmospheric character of a situation" (Tellenbach 1968: 41). The atmosphere is hence not an extra-sensual dimension of the situation but is a productive force that moulds sensorial regimes. Situation-listening then is, in a second sense, an auditory mode of existence evoked by the atmospheric dynamics of a situation, dynamics that always seize the felt-body as an indivisible whole. Listening in the atmosphere is an emergent mode of existence rather than a modal means of immersion. Such listening is hence no longer to be confused with the physiological sense of hearing as mediating between self and world. Standing at the village square in Manna we became listeners by way of a diffuse non-personal yet telling tension that had condensed in music, sound and silence as a powerful atmosphere. As crowds of villagers stood in silence observing the rhythm-making of musicians in their midst, listening was watching, and watching was listening.

Here is a first moment of segmentation. Crucially, Schmitz emphasises that the diffusity of situations is internal, "they are set off against the outside and hang together internally" (2009: 47). Atmosphere as situation here is not merely undifferentiated mood, but atmosphere condenses in sensual regions and in telling impressions that may be vague regarding their constituents and identity but that are nevertheless particular regarding their state of affair and trajectory. The situation at the village square was charged with a particular tension precisely because it stood in suggestive contrast to the holiday where the very same place would reverberate to music and to the dancing of men and women all dressed up to the nines. AiGianNan explains this, by comparing situations of music-making, drawing a link between music performance and an atmosphere of shame:

> Right now it is very difficult to sing. It is daytime. In the daytime I am very much ashamed [of singing]. [But] in the evening it will be just about right. . . . It is very uncommon to sing inside the house during the daytime.

Yet, the question remains how a person relates to, or is related to, atmospheric dynamics of music. And how can this relation be analysed if not by an antecedent constellation and hence singularisation of bodies? Furthermore, how can anthropology or musicology account for the atmospheric dimensions of a situation if such situations are undividable wholes at which no individual meanings and discrete bodies can be isolated? When I listed different elements of the situation on the village square – the crowd, the music, the beat of the drum – did my analysis not at least tend to fall back into the rationale of constellationalism? To follow up on these matters I will have to turn to the rather notorious problem of subject and object in atmosphere studies to unpack how Schmitz reintroduces difference and singularity that are the very precondition to being a subject.

Atmospheric dynamics: contraction and expansion

One common understanding of atmosphere or ambience is simply to identify it with that which "surrounds us" (Friberg 2012) or that which envelops a subject. This topological, centric and dichotomous constellation of subject and environment is problematic, because one still lacks an explanation of how atmosphere gets into the subject or how the subject gets into atmosphere. However, one may pose this question differently: how does atmosphere come to appear as a surrounding? For Böhme, atmospheres are instead phenomena of in-betweenness, precisely between subject and object (Böhme 2013). He writes that "atmosphere itself is not a thing but rather a floating *In-between*, in between the things and the perceiving subjects" (ibid., 107). Again, such a notion of the in-between is incompatible with Schmitz's understanding as it suggests that atmosphere here is "merely a bridge between two given entities" (Schmitz 2005: 273). Conversely, the subject is, for Schmitz, not naturally given; person and world are not initially divided,[9] and there is no space between them that atmosphere could occupy. Rather, it is precisely this axiomatic division that Schmitz aims to subvert by way of turning the subject's inner-sphere, the territory of feelings, inside out, making the very distinction between internal and external space untenable (Schmitz [1969] 2005: 86). In doing so, Schmitz pre-empts what, much later, Morton (2002: 54) suggests doing: to "deconstruct personhood into ambience, atmosphere, surroundings, dwelling, environment".[10] The question of how a subject immerses herself into an atmosphere, as Goodman (2010: xiv) phrases it, or opens her (inner) self up to an atmosphere, both of which are topological metaphors of atmospheric involvement, are, so Schmitz reminds us, misleading (Schmitz [1969] 2005, 12).[11] Instead "everything is atmospheric" (Griffero 2014: 143), there is no outside, no transcendental Archimedean point of access (Bollnow 1941: 134), no extra-atmospheric stability no "solid to hang onto" (Morton 2002: 54). This "everything" is however not a quantifiable dimension of atmosphere but is part of its transgressive force and authority to claim excessive presence and to involve everyone and everything.

Dismissing the secluded subject as epistemic locus, Schmitz posits the felt-body (*Leib*) as bedrock of his thinking.[12] The felt-body is, for Schmitz, neither physical materiality nor a regime of senses (Schmitz [1978] 2015:17); its very essence is, rather, *dynamism*. This dynamism of the felt-body centres in an axis Schmitz terms vital drive (*vitaler Antrieb*) which unfolds in the "intertwined tendencies of contraction and expansion", of "tension and swelling" (Schmitz 2009: 34). Since the felt-body is not a secluded entity but rather is an organic bundle of dynamics, it is structurally open to situational dynamics. Crucially, sound, music[13] and atmosphere are too essentially dynamism – they all take shape in expansion and contraction, tension and swelling, and operate by way of stirring and movement suggestions (*Bewegungssuggestionen*) (Schmitz [1978] 2005: 245).[14] Dynamism is, in short, the shared sphere of felt-body, music and atmosphere; in their structural equality they are indivisibly entangled and congeneric (Schmitz 2014a: 85). Ensuing from the felt-body, New Phenomenology seems naturally anthropocentric.[15] Yet, the way a human person is, for Schmitz, always already caught up in atmospheric dynamics allows for a rather ecological perspective on atmosphere. In this perspective, feelings that are atmospheres firstly cannot be reduced to a narrow notion of human perception let alone an anthropocentric introjection of feelings, and secondly nonhuman forces, movement suggestions, and environmental dynamics need to be concidered as being part of a person's felt existence.

With the emphasis on vitality, New Phenomenology grants ontological priority to dynamism over the singularity and identity of things and subjects and their relations; dynamism here in fact precedes the very difference between relations and relata. As a consequence, emphasising the union of the perceived and the act of perception in Schmitz' writings, the fundamental mode of perception is and moreover being, for him, encorporation (*Einleibung*), a dynamic process of material synergies that spawn modes of existence.

Subject-moments

And here is where singularity and difference reappear. One of these potential modes of existence is subjecthood or what Schmitz more aptly terms personhood (*Personsein*). Personhood gradually takes shape in a process of "personal emancipation"[16] out of the dynamic stirrings and the diffuse meaningfulness of situations. This emancipation occurs with the increase of contrast between the own and the foreign, by splitting the dynamism into cause and effect, by separating subject and object. A personal situation arises that develops and transforms lifelong in processes of personal emancipation (Schmitz 2015: 50). These processes fundamentally depend on, and progressively amplify, difference and singularisation in order for the personality to solidify and to develop. Instead of assuming a substantially determinable subject, New Phenomenology enables us to focus on *situational subjects*, or what Schmitz terms "subject-moments" (*Subjektmoment*) (Schmitz [1969] 2005: 89).[17] Schmitz here empties the subject of a topological and identitarian logic: while in the spatial notion of atmosphere, singularity – the very precondition of the existence of a subject – is linked to position and place, in Schmitz's situationist approach singularity is linked to a moment.

Methodology of atmosphere

From this can be drawn two methodological inferences: firstly, due to the affinity of atmosphere and vital drive, "atmosphere grips the vital drive" (Schmitz 2009: 86) and not just an already constituted subject. In doing so atmosphere has an immediate effect on the link between the diffuse meaningfulness and the emergence of subject-moments, that is, on the very foundation of the becoming-person (Schmitz 2009: 86). The atmosphere, then, as Tellenbach writes "constitutes a shift in the *Dasein*, investing it with a tone before one can talk in a positive or negative sense about new meaningfulness or even meanings" (Tellenbach 1968: 73). Atmospheres in this perspective are not mere qualities (of objects or spaces) but shifts, stirrings, interventions, alterations, in the dynamism of the felt-body, operating at the situational threshold between unsegmented diffuse meaningfulness and the segmented situation of emancipated personhood. If atmospheres are not the object of perception for a transcendental subject but rather have the potential to impact on the vital drive, the very foundation of the person, then the atmospheric dynamics, so I shall argue, can be inferred from their effects and traces. After all, since cause and mode of influence coincide, atmospheres *are* their effects. Secondly, due to music's idiosyncratic capacity to evoke stirrings in the felt-body, music becomes a prime milieu for the constitution of atmospheres (Schmitz [1978] 2005: 260).

It is only by way of personal emancipation that a person comes to relate to a situational atmosphere, that is puts herself into a relation to something whether by way of succumbing and/or resisting to the atmosphere in which she is always already entangled. The emancipation from the stirrings of atmosphere, the process in the course of which the unsegmented whole disperses into difference, is done by way of operations of segmentation and interventions. The most salient example Schmitz gives here is language, an operation of identifying different objects that are put in relation to each other by means of grammar. But there are subtler and more indirect operations of emancipation. While Schmitz contends that atmospheres cannot be produced from scratch – his most prominent point of disagreement with Böhme[18] – he notes that atmospheres can be cultured, tempered and curbed (Schmitz 2014a: 29). Such operations of empowerment and of personal emancipation, neutralisation and differentiation can again indirectly inform about the vectorial dynamics, intensities, trajectories or modes and rhythms of contraction and expansion of the atmosphere. Atmospheres here are neither primarily related to constellations, objects or spaces, nor merely to the dynamics of the felt body – and here I propose going with Schmitz beyond Schmitz – but rather atmospheres are embedded in situational process and temporal *performance*. Jan Assmann makes this point when he writes that, "the source of an atmospheric emanation is not embodied in a present person, much rather atmospheres are created in performance [*im Vollzug*]" (Assmann 1989). Bringing this together with Schmitz's notion of dynamics and suggestions of movements we can extend Assmann's point by saying that atmospheres are *co*-created in performance.

Far from simply aiming to describe or represent an atmosphere in an ethnographic account, I suggest that to consider the atmospheric dimensions of a situation means to trace the dynamism that permeates music, atmosphere, and felt-body in performance. This requires investigating the operations of empowerment and personal emancipation, and to ask how atmospheric dynamics are inscribed into performance and situational-subjects. In the following, before going into the temporal dimension which the notion of performance hints to, I shall first focus on particularly telling operations and performances that point to the musical dynamics of atmosphere in Manna.

Tempering and curbing atmospheres and the dynamism of voice

When a feast or a dinner drew to a close in Manna, a group of people sitting in a circle around a table in the stilt house would often start to challenge each other in song and in doing so evoke a situation marked by an atmosphere of timidity and shame. Shame became manifest in the lowering of the gaze and in withdrawing oneself from the group, sometimes literally hiding in the dark background when one was challenged to sing. This seems to affirm the trajectory of shame as an overly severe contraction of the vital drive, preventing an unfolding of the felt-body. Yet the dynamic play of unfolding and contraction of tension and swelling of the felt-body were needed for singing: to open ones' mouth, tense the vocal cords, – indeed most of the songs sung by men were intonated in a reedy head voice solely used in singing – inhale, hold the breath, to anticipatie the hearing of ones' own resounding voice and toembrace the musical suggestions of movement in song. In short, the situation of an appointed singer was precarious and only further contributed to an at times overwhelming tension. Thus, when a person was summoned to sing, a whole range of operations was needed to take control over the contracting atmosphere of timidity that inhibited the dynamism needed for singing. Here, I shall identify three such operations that provide insight into the atmospheric dynamics and tensions at play in situations of communal music-making in Manna.

A first means of controlling feelings of shame was intoxication, in which the person asked to sing was given alcohol. Not only was alcohol repeatedly poured into his[19] cup, but he was verbally encouraged to drink and to sing even to the degree that someone would raise the cup of hard liquor to his mouth and literally force him to drink, force him to subdue the feeling of timidity that had overpowered him and that prevented him from singing. The effect of alcohol consumption before singing was expressed in the figure of speech "to brush one's teeth". This operation of opening, cleaning, moving, spitting, and swallowing in brushing the teeth hints to the preparation of mouth, face, saliva, jaw, glottis and throat for singing. Since alcohol acts as a relaxant of muscles, and moreover as a holistic alteration of the felt-body towards its expansion, the person becomes poised to engage in musical dynamics. Affecting the vital drive such intoxication would free a person from the grip of shame.

The associated dimensions in Manna of drinking alcohol, namely cleaning, reveal another implication of this operation. The consumption of liquid, whether alcoholic or not, had an effect on the quality of voice – more precisely on timbre. Drinking potentially manipulated timbre by eliminating grain in the voice, by cleaning the voice of roughness. In his work on environmental aesthetics, Timothy Morton (2007) instances the timbral as one of the most salient aspects of atmosphere.[20] Timbre foregrounds the voices itself, it "evoke[s] the medium that utters them" (ibid. 40), the singer. At the same time, timbre belongs to the medium of utterance. Ultimately, timbre is always both a reference to the singer and the medium of song. Thus, following Morton, timbre undermines the distinction between background and foreground; it elicits, in Schmitz words, a diffuse meaningfulness. The cleaning of the voice and the concomitant reduction of personal timbre however worked against this subversive force of the timbral and enabled the withdrawing of the singer into a medial position while foregrounding song. In controlling timbre by way of cleaning the throat, the singer was able to exercise some control over the dynamics of atmosphere which precisely threatened to blu the distinction between individual and medium, between singer and voice.

A second operation was the handing of a sheet of paper or banana leaf to the appointed singer for her to hold onto while singing. It was explained by a musician that "this is because when one sings one is a little shy, this paper works like this: when you hold it [holds it in front of her upper body], you are not shy anymore". Whereas the contraction of inhibition and shame elicited an urge to disappear, the dynamics of voice were one of appearance, a presence that exceeded the corporeal visible so far as the sound of the voice permeated into the beyond of the wooden stilt house. Singing meant a radically expanded presence. Torn between these contradictory tendencies of the wish to disappear and the inevitable, even transgressive presence in singing, the leaf enabled the singer to take control of the situation. Holding it in front of her face or body she was able to forge a secluded space between herself and the leaf, thus interfering with and shaping the dynamics of her felt-body. The private space created by the leaf potentially ensured just enough intimacy as was needed for the existential subversion released in singing. The paper worked precisely as what psychoanalyst Donald Winnicott would identify as comfort object, a transitional object of a becoming self. Herein, the paper was both a means of visual or felt disappearance; a way of distinguishing a person from the group, marking her as singer before even starting to sing: a singularisation, a differentiation between self and other, a personal emancipation, a constellation. Schmitz suggests that "a person can at the same time be at more than one level of personal emancipation" (Schmitz 2009: 96), that is, both entangled in the diffuse meaningfulness of atmosphere and simultaneously in a mode of emancipated selfhood through an immediate identification with the pervasive sound of one's own voice, a radically transformed presence.

A third operation for taking control over the contraction that seized a singer concerned a melodic gambit: the player of the lute would reiterate a musical phrase over and over again, a luring loop. The repetitive movement-sound of alternating intervals, often octaves and fifth, would, by means of the musical suggestions of movement, affect the vital drive of the appointed singer, previously locked in a node of contraction due to the feeling of shame. The suggestions of movement as nfurled by the lute would potentially loosen the singer's contraction. The singer ld then seize the expansion as sketched by the repetitious movement-sound, forces with the lute, launch into song, and, while swaying back and forth, orate and enact the musical dynamism. The liminal moment of joining-in is d in the musical texture of the song: The melodic movement intonated by r clings to the patterned movement of the lute. As the singer is in unison ute on the first or last stressed beat of the opening phrase, the dynamism ion and expansion in lute and voice coalesce. While lute-playing was encumbered with feelings of shame as was singing, the lute assisted an atmosphere for singing. This musical operation worked as an rce precisely because it aimed at seducing the singer into song, of nto the musical trajectory, a situational atmosphere, irreducible to of voice and lute. Even if the instrument player and singer were person this musical operation was efficacious in situations of lity.

These very particular operations of personal emancipation in situations of communal music-making illustrate that atmospheric dynamics are not determining every action, position or movement. Neither are they reducible to a spatial constellation. On the contrary, performance, in fact collective performance, proves capable of massively altering both atmosphere and modes of presence, launching and unfolding continuous atmospheric shifts: from shame to the pride and joy taken in singing, from bashfulness to audacity. Put another way, these operations elucidate what Andreas Philippopoulos-Mihalopoulos (2014: 124) has identified as the "main atmospheric paradox: atmosphere is always both engineered and emergent. Bodies constitute atmosphere and atmosphere constitutes bodies". Causality is indeed dubious when it comes to atmosphere. I suggest going one step further and argue that atmosphere eludes *any* causal constellation. Rather, I contend, it is this very tension between being always already entangled in compelling atmospheric dynamics on the one hand and the potential operations of personal emancipation and of altering atmosphere on the other, which constitutes their generic force, where atmospheric dynamics come to be inscribed in processes of selves and collectives. And it is here that temporality is no longer ancillary but is an intrinsic and texturing dimension of atmosphere.

Duration of atmosphere

> There is [. . .] a duration peculiar to the musical object. This duration is always movement, a movement which includes the movement of a soul fascinated by sound and immersed in a certain atmosphere.
>
> (Dufrenne 1973: 184)

Bille et al. have pointed to the "temporality of atmospheric encounters" and to the "multi-temporal tension" of atmospheres that are "at the same time a product of the past and future" (Bille et al. 2015: 34). But there is something more to atmospheric temporality than their change over time and the implicit historicity that Bille et al. argue to be contingent upon changes of environment, human values and cultural premises (ibid.). Along the lines of Morton (2007: 166), there is an atmospheric difference if one listens to a long musical phrase or entire piece as opposed to a short phrase: "the 'same' atmosphere is never the 'same' as itself". Temporality here is a fundamental dimension of how atmosphere differentiates. Temporality itself is a generic force. Schmitz accounts for this temporal dimension of atmosphere with his notion of duration.[21] Both sound and music produce and shape duration by amalgamating intensive[22] length and density of duration. These intensive musical temporalities are, for Schmitz, manifest in temporal dimensions of extension and contraction such as in the extension or eventual contraction as may be suggested in a long tone or a sprawling interval, and hence take shape as feelings that are atmospheres (Schmitz 2014b: 61). While Schmitz determines how temporal dynamics of music relate to dynamics in the felt-body such as density of duration with contraction, I refrain here from such universalising mappings. However, the link between duration and contraction and expansion is crucial. Musical duration is neither translated into what has been termed emotions nor does it signify a certain feeling. Rather just as the modality of music and felt-body is dynamism, as I had argued before, they too share the temporalities that unfold in the dynamics of contraction and expansion. Music does not induce a certain feeling but in the shared situational duration, shaped by the tendencies of contraction and expansion, both music and feelings unfold in indivisible correlation. I want to suggest here that this very unfolding *is* atmosphere – the shifting in *Dasein* that Tellenbach alludes to – the performance, transforming, transgressing, transposing, altering and the being-moved.

When a singer in Manna enters the repetitious musical movement of the lute, a melody arises that fashions the timing of community. Listeners often sit in silence and, while listening, are captivated by the sequence of musical time. Since sung phrases can vary in length one is not entirely sure about the kind of duration to expect; temporal tension swells. At some point listening is waiting. As the singer continues to loop a musical gesture whose tonal material lies just beneath the fundamental, listeners' attention strains towards that fundamental, agog for release. At some point singing is waiting. A waiting that resonates with the listener awaiting the release. At some point singing is listening. Duration and tension are mutually constitutive precisely because atmospheric time is unsegmented and indivisible. The tension wanes as the song returns again to the tonic: a deceleration, sustained pitches, a high ascending tonal movement in the voice and a tremolo in the lute. Ovations by the listeners at the end of particularly long phrases and at the end of the song mark the moment where intensive duration has come to a peak and inevitably collapses. The release, manifest in vocal acclamations and in the relaxation and moving of the listening body, reveals the tension in song. Agglomerating temporal thresholds, swelling into distinctive durations, music brings atmosphere into appearance as diverse situations. Short repetitive pitches throughout the verse of the song and elongated tones towards its end can summon entirely different dynamics of contraction and expansion and hence shift and alter atmosphere. After all, contraction and expansion conflate as rhythm. As the repetitious movement-sound of the lute forms a pulsating rhythm that penetrates the situation, a collective is formed in the musical duration. This does not mean a radical determinism. Rather, as the operations of emancipation in Manna exemplify, the dynamism of atmosphere is inscribed both in the succumbing to its striving force and in the resistance to it, in short, in collective performance.

Conclusion: towards a concept of atmosphere

Atmosphere subverts the distinction between self and world, inside and outside, between medium and thing, background and foreground. Individuality and difference are consumed in the situational (and not just spatial) meaningfulness of the atmosphere. Subversive harmony. A form of collective performance, feelings that are not centred on human experience: it escapes the very idea of a centre (Morton 2002: 52). Yet, atmosphere is not merely the amorphous flow of matter and becoming. In fact, when Schmitz defines atmosphere both as diffuse meaningfulness *and* as half-thing (*Halbding*),[23] that is, as an entity with its own particular form and manner of appearance, he indicates the simultaneity of holism and processes of segmentation in regard to atmosphere. As the examples of music-making and feelings of shame in Manna elucidate, atmosphere cannot be separated from the bodies and forces present. Yet these are not merely the material basis from which atmosphere emerges. It is not the relation between bodies or subjects and objects that is relevant but it is the continuous changing relation between the unsegmented diffuse meaningfulness and subject-moments. Atmosphere, as I have argued, is inscribed precisely in the oscillation between segmented constellations and internally-diffuse situations by both subverting identity and difference and by provoking processes of emancipation. As cause and mode of influence conflate, what atmosphere *does* and what it *is* collapses. Atmosphere is in short the very alteration it induces. Music as atmosphere is the dynamism that it unfurls, the "shifts in *Dasein*", the being lifted up, being moved, finding oneself in nostalgic memories of a lover, or the striving against feelings of shame. Hence I suggest that in tracing atmosphere, an engagement with its topology will only get us so far. A focus on duration and dynamics will instead enable us to sound out atmosphere as process and performance (*Vollzug*).

Constellations of singular objects and worlds segmented along the lines of senses cannot mark the beginning of a study of atmospheres. Dissecting a situation into allegedly constitutive parts is in fact, as Schmitz emphasises, an existential human act to take control over atmosphere and diffuse its authority. He writes, "humans [as opposed to animals] can take charge of and reconstruct the situations [. . .] in such a manner so as to implement hierarchy and wield authority" (Schmitz 2005: 9). And it is here that the notion of atmosphere may prove to be of great potential for both anthropology and musicology. As atmosphere is an operative force that mediates between self and collective shaping the very relation between processes of individuation and collective, atmosphere as situation could serve as a primary research category, as a unit of anthropological enquiry that precedes the constitution of social group, sub-culture, ethnic identity or discourse community. This is precisely because it enables us to ask how difference and segmentation arise and vanishes. Furthermore, it is precisely the ambiguity and vagueness of atmosphere as collective feeling that affords great generative conceptual potential in rethinking the collective as situational process and the ever shifting, and non-unitary subject as subject-moment.

Notes

1 The notion of "New Phenomenology" only emerges after Schmitz' completion of the *System der Philosophie* in 1980. For an introduction see Schmitz (2009).
2 While those theories of atmosphere that are often labelled ambience proceed from a notion of atmosphere as qualities of surrounding space (Böhme 2013: 23) or environment (Morton 2007), Schmitz concept of atmosphere is at its heart a theory of feelings.
3 I would like to thank Ruard Absaroka, Bernd Bösel, Mark Porter and the editors of this volume for their comments on earlier drafts of this paper. I am also indebted to Hermann Schmitz for his generous discussions of the arguments in this paper.

4 Timothy Morton, who also associates himself with objectoriented ontology counterbalances this implication of singularity with his notion of ambient poetics.
5 Since, for Schmitz, "classical ontology" and its implied rationality imagines their world as a synchronic constellation of singular elements (Schmitz 1998: 185), atmosphere naturally exceeds rational description.
6 Both animals and humans are caught up in situations, yet their ways of dealing with situations are radically different since humans dissect situations via language, so Schmitz argues.
7 This too makes clear that it would abridge and distort Schmitz' argument if his notion of atmosphere merely referred to a certain aspect of the world, namely, as it does for Böhme, aesthetics.
8 This does not mean that music is reducible to atmosphere nor that atmosphere is reducible to music.
9 This is of course the classic tenet of phenomenology.
10 Note that Schmitz highly differentiated notion of personhood (*Personsein*) is not to be confused with Morton's notion of personhood.
11 Schmitz ([1969] 2005: 12) writes "Die Berechtigung der räumlichen Bilder vom Insichsein, Inhalt, Heraustreten, Hinausgreifen pp. wird nicht begründet".
12 In German debates, Schmitz is commonly identified as *Leibphilosoph* (philosopher of the felt-body).
13 Schmitz explicitly names both sound as acoustic phenomenon and music such as hymns, fugues or breakdance.
14 While these dualist terms of a dynamic and movement seem reductive regarding the multiplicity of musical process, movement and dynamic Schmitz conceives of them very broadly to accommodate the many different dynamics at play in music
15 Or at least is it centred on sentient beings.
16 Not an emancipation that is personal but an emancipation of the person as person.
17 Crucially, atmosphere is in this view not (just) a pre-subjective force. Atmosphere preconditions and is inscribed in the very constitution of difference between bodies, subjectivities, objects (Schmitz 1998: 186).
18 The paradigm of creating an atmosphere is, for Böhme, a scene designer who creates the atmosphere of a stage for a particular scene in a play (Böhme 2013).
19 Songs were sung both by women and men but the drinking of alcohol was widely restricted to men. Women engaged the same drinking practice; however they would replace alcohol with juice.
20 Note that Morton chooses the term ambience over atmosphere.
21 This has of course, as Schmitz admits, previously been elaborated by Henri Bergson. Bergson however, according to Schmitz, fails to explain how intensive duration and extensive time or singular moments are related (Schmitz 2014b: 261).
22 Schmitz uses the term intensive to refer to an expansion, here length, that is not metered.
23 Half-things differ from full-things in two ways. Their duration can be interrupted, that is, they come and go, without there being any point in asking what they did in the meantime, and their causality is immediate in that cause and influence overlap. A typical example is the voice of a human.

Bibliography

Anderson, B. 2009. Affective Atmospheres. *Emotion, Space and Society* 2, 77–81.

Assmann, J. 1989. Der schöne Tag. Sinnlichkeit und Vergänglichkeit im altägyptischen Fest. In *Das Fest*, edited by W. Haug and R. Warning. München: Fink, 3–28.

Bille, M., Bjerregaard, P. and Sørensen, T. F. 2015. Staging Atmospheres. Materiality, Culture, and the Texture of the In-Between. *Emotion, Space and Society* 15, 31–38.

Böhme, G. 1993. Atmosphere as the Fundamental Concept of a New Aesthetics. In Thesis Eleven 36 (1), 113–126.

Böhme, G. 2013. *Atmosphäre: Essays zur neuen Ästhetik*. Berlin: Suhrkamp.

Bollnow, O. F. 1941. *Das Wesen der Stimmungen*. Frankfurt am Main: V. Klostermann.

Clayton, M. 2008. Toward an Ethnomusicology of Sound Experience. In *The New (Ethno) musicologies*, edited by H. Stobart. Lanham, MD: Scarecrow Press, 135–169.

Connor, S. 2006. *Atmospherics* [Online]. Available at http://stevenconnor.com/atmospherics/atmospherics.pdf [Accessed 5 April 2016].

Cook, N. 2014. *Beyond the Score: Music as Performance*. Oxford: Oxford University Press.

Dufrenne, M. 1973. *The Phenomenology of Aesthetic Experience*. Evanston: Northwestern University Press.

Ford, T. H. 2013. Aura in the Anthropocene. *symploke* 21(1), 65–82.

Friberg, C. 2012. Hermeneutics of Ambiance. In *Ambiances in Action: Proceedings of the 2nd International Congress on Ambiances*, edited by J.-P. Thibeaud and D. Siret. Montreal: International Ambence Network, 671–676.

Griffero, T. 2014. Atmospheres: Aesthetics of Emotional Spaces. Farnham: Ashgate.

Goodman, S. 2010. *Sonic Warfare: Sound, Affect, and the Ecology of Fear*. Cambridge, MA: MIT Press.

Gumbrecht, H. U. 2011. *Stimmungen lesen: Über eine verdeckte Wirklichkeit der Literatur*. Edition Akzente. München: Hanser.

Harman, G. 2011. *The Quadruple Object*. Winchester: Zero Books.

Heidegger, M. 2010. *Being and Time*. Albany: State University of New York Press.

Morton, T. 2002. Why Ambient Poetics? Outline for a Depthless Ecology. *The Wordsworth Circle* 33(1), 52–56.

Morton, T. 2007. *Ecology Without Nature: Rethinking Environmental Aesthetics*. Cambridge, MA: Harvard University Press.

Philippopoulos-Mihalopoulos, A. 2014. *Spatial Justice: Body, Lawscape, Atmosphere*. Abingdon: Taylor & Francis.

Riedel, F. 2015. Music as Atmosphere: Lines of Becoming in Congregational Worship. *Lebenswelt* 6, 80–111.

Schmitz, H. [1969] 2005. *Der Gefühlsraum*. Bonn: Bouvier.

Schmitz, H. [1978] 2005. *Die Wahrnehmung*. Bonn: Bouvier.

Schmitz, H. 1998. Situationen und Atmosphären. Zur Ästhetik und Ontologie bei Gernot Böhme. In *Naturerkenntnis und Natursein: Für Gernot Böhme*, edited by G. Böhme, M. Hauskeller, C. Rehmann-Sutter and G. Schiemann. Frankfurt am Main: Suhrkamp, 176–190.

Schmitz, H. 2005. *Situationen und Konstellationen: Wider die Ideologie totaler Vernetzung*. Freiburg: Verlag K. Alber.

Schmitz, H. 2009. *Kurze Einführung in die Neue Phänomenologie*. Freiburg im Breisgau: Verlag K. Alber.

Schmitz, H. 2011. Der Leib.

Schmitz, H. 2014a. *Atmosphären*. Freiburg: Alber.

Schmitz, H. 2014b. *Phänomenologie der Zeit*. Freiburg im Breisgau: Verlag K. Alber.

Schmitz, H. 2015. *selbst sein: Über Identität, Subjektivität und Personalität*. Freiburg im Breisgau: Alber.

Schmitz, H., Müllan R. O. and Slaby, J. 2011. Emotions Outside the Box—The New Phenomenology of Feeling and Corporeality. Phenom Cogn Sci 10 (2), 241–259.

Tellenbach, H. 1968. *Geschmack und Atmophäre: Medien menschlichen Elementarkontaktes*. Salzburg: Otto Müller Verlag.

Index

For Product Safety Concerns and Information please contact our EU
representative GPSR@taylorandfrancis.com
Taylor & Francis Verlag GmbH, Kaufingerstraße 24, 80331 München, Germany

www.ingramcontent.com/pod-product-compliance
Ingram Content Group UK Ltd.
Pitfield, Milton Keynes, MK11 3LW, UK
UKHW022126190926
475702UK00013B/1676
* 9 7 8 0 3 6 7 5 9 3 6 6 7 *